CHILTON'S GUIDE TO
TURBOCHARGED CARS and TRUCKS

Mike Stone

v-14
#990

Vice President and General Manager JOHN P. KUSHNERICK
Managing Editor KERRY A. FREEMAN, S.A.E.
Senior Editor RICHARD J. RIVELE, S.A.E.
Editor JOHN M. BAXTER

CHILTON BOOK COMPANY
Radnor, Pennsylvania
19089

Special thanks go to my wife, Toni Stone, for her help and emotional boost in writing this book. Thanks also to TMS Word Processing Services for their word processing and to Beck Beckwith for his help in writing Chapter V. And I am compelled to thank those in the industry who enriched this book with their information and photographs.

Manufactured in the United States of America
1234567890   3210987654

Chilton's Guide to Turbocharged Cars and Trucks
ISBN 0-8019-7397-X pbk.
Library of Congress Catalog Card No. 83-45309

14 75

Contents

910234

1
Why Turbocharge?

Why a Turbo? The obvious answer is power. Power for passing, power for acceleration, power for top speed, power for climbing hills, and power for pulling heavy loads.

If there's one thing this country needs more of, it's acceleration. The doddering Darts and crawling Chevettes that clog our back roads into slow trains of traffic make frequent full throttle passes a necessity, if you want to maintain a reasonable speed. These slow trains of traffic are created by drivers who lack an enthusiasm for passing. Yet all but the dullest of deadbeats must occasionally think, "I'd like to blow this roadhogging slowpoke off the road." Most people just don't have the guts to do it.

That's partly because they also don't have the horsepower. The average car takes 16 seconds to get from 0 to 60 miles per hour. The Chevette diesel takes more than 21 seconds! That kind of leisurely speed increase just won't allow safe backroad passes. It takes sub-10–second 0 to 60s to pull off these acts of aggression.

A quick car can also accelerate off entrance ramps to safely enter the flow of highway traffic, and rapidly accelerate back up to cruising speed when slow traffic finally gets out of the way. And to the speed freak, there's nothing like the thrust of a quick car rocketing from turn to turn on a favorite section of back road.

Turbocharging increases horsepower, allowing safe backroad passes.

Horsepower, as measured on a dynamometer, is the result of displacement, engine speed, and mean cylinder pressures.

0 to 60 mph acceleration is almost entirely a function of horsepower versus weight. The more horsepower a car has, and the less weight, the quicker it will accelerate. Aerodynamics, tires, transmissions, axle ratios, and limited slips all play small but significant parts in 0 to 60 mph acceleration, but horsepower and weight are by far the prime factors.

As anyone who has ever lugged a loaded truck or trailer up I-70 into Denver knows, it takes lots of horsepower to pull a heavy load or climb long, steep hills at a decent rate. Top speed is almost entirely a factor of horsepower and aerodynamic drag. The more horsepower a vehicle has, and the less wind pressure against the vehicle, the faster it can go.

So horsepower is of critical importance for acceleration, top speed, hill climbing, and pulling heavy loads.

There are three factors which determine horsepower. The first is displacement—the total volume (cubic inches or cubic centimeters) of the engine. The second factor is r.p.m., or the speed (revolutions per minute) that the crankshaft turns. Of final importance is the mean effective pressure (m.e.p.) in the cylinder, which is determined by the design of carburetion, manifolding, cylinderhead porting, valves, camshafts, compression ratio, and ignition timing, combustion chamber design, fuel type, supercharging, and turbocharging. M.e.p. is a complex concept, but you can readily visualize that it relates almost directly to the amount of fuel and air in the cylinder and the pressure they generate during the downstroke. In fact, the equation for horsepower is as simple as this: Horsepower = rpm × displacement × m.e.p.

But power isn't the whole reason for turbocharging. After all, as the equation for horsepower shows, there are other, simpler ways to increase power. Displacement, for instance, plays a prominent part in our equation for horsepower, and increasing an engine's displacement or using a bigger engine is certainly an effective way to increase power. But a bigger engine uses more fuel, and is usually heavier, hurting braking performance, handling, etc.

Conventional four-stroke gasoline engine modifications, such as high performance carburetion systems and camshafts, higher compression ratios, and emissions equipment removal raise mean effective pressure and raise the engine speed at which the engine can produce power. In the case of emissions air pump removal, horsepower-wasting air pump drag is eliminated. When properly orchestrated (most backyard emissions system tinkering causes more harm than good), these normally-aspirated modifications frequently yield a 20 to 40% increase in horsepower while improving gas mileage.

The amount of this power increase varies so greatly because the restrictions, tuning, and emissions equipment of different cars varies greatly. When normally aspirated modifications get too radical in the quest for peak horsepower, low- and mid-rpm horsepower goes down the tubes, and fuel economy goes with it. This loss of low and mid range power is the result of high overlap camshafts, big carburetors, and cylinder head porting which sacrifice good low and mid-range intake and exhaust flow for the sake of improving high rpm flow. Of course, most of these conventional modifications have an additional drawback: They increase exhaust emissions, as tested by the Environmental Protection Agency. And that's one drawback the auto manufactures can't get around.

Of the 190 normally aspirated new cars for sale in the United States, only 20 can get to 60 miles per hour in less than 10 seconds. Virtually all of the normally aspirated domestic cars which pass this hurdle are V-8s, getting no more than 18 EPA city miles per gallon. The average base price of normally aspirated imported cars with sub-10 second 0–60 mph acceleration is $24,000. All but a couple of the 37 turbocharged gasoline cars for sale in the U.S. since 1980 get to 60 mph mph in less than 10 seconds— this at an average EPA city fuel economy of 22 mpg and an average base price of $14,000.

So when you look at the whole picture: horsepower, fuel economy, exhaust emissions, engine size and weight, turbocharging is hard to beat. At this point it's not the only way to go, but one very good way. The normally-aspirated Dodge Charger 2.2,

Four cylinder turbocharged gasoline engines, like this Ford 1.6 liter Escort/EXP engine, offer a combination of power, weight, size, and fuel efficiency that's hard to beat.

for instance, gets 28 EPA city mpg, while pulling 9.6 second 0 to 60s, a combination which can only be beaten by the Ford Escort Turbo GT.

Normally-aspirated cars have traditionally had some advantages over turbos. Many early turbocars had poor throttle response, so when you put your foot down you had to wait a half second or longer before you'd get power. But with most modern turbos, the one-tenth of a second loss in throttle response is almost imperceptible. The danger of engine detonation is always present with turbos, but modern electronic controls have minimized the chances. However, on a high mileage turbo, wear, corrosion, dirt, or faulty controls are more likely to cause the death-rattle of detonation than on a normally aspirated engine. Burnt exhaust valves and blown gaskets are problems many turbokit owners have experienced, but upgraded materials and designs make these problems rare on factory turbo cars. Expensive premium unleaded fuel is the recommended diet for most turbos, though low 87 octane unleaded can be used in most factory turbo cars with a slight power and fuel economy loss. And in some cases, turbos require more frequent oil changes than normally aspirated engines.

Some of today's high tech turbo systems and components, like electronically-controlled wastegates and intercoolers, are making the power outputs and efficiency of turbocharged engines so great that turbos may be the only way to get the best combination of fuel economy and horsepower in the near future. An excellent example is the Mustang SVO, which pulls a fantastic 174 horsepower out of its 2.3 liter four cylinder turbo, and gets 21 EPA city mpg, despite its somewhat overweight, oversized chassis, which was originally designed to accommodate a V-8.

This discussion has thus far ignored turbo diesels. Turbos bring the otherwise sluggardly diesels up to average, or in some cases better than average, acceleration standards for gasoline cars. The quickest turbo diesels still take more than twelve seconds to reach 60 mph from a standstill, so no turbo diesels can be classified as "quick." Turbocharged diesels suffer little or no fuel economy penality when compared with normally aspirated diesels, and may actually offer improved economy, so if you're into diesels, you might as well go turbo. However, today's four cylinder gasoline-powered cars get such good gas milage, it's difficult to justify a diesel—unless you save money by illicitly using home heating oil as a fuel. In fact, the best EPA city fuel mileage car for '84 is the gasoline-powered Honda CRX, which gets better EPA city mileage than any diesel. The one thing diesel power may do for the car buyer is permit him to avoid downsizing his vehicle while improving mileage by a modest percentage.

On the other end of the automotive spectrum are those all-out power fiends with V-

Pontiac's 2000 Turbo uses an electronically-controlled wastegate to obtain 150 horsepower out of 1.8 liters—more power per liter than any other production car.

Turbo diesels like this Volkswagen Jetta offer excellent fuel economy with average (by gasoline car standards) acceleration.

8 Mustangs, Camaros, Firebirds, Corvettes, etc., who want still more power. Once they've benefitted from the common hot rod modifications, they can either get more radical with cams and carbs, which will sacrifice mid-range power for more high rpm power, try nitrous oxide injection which is only good for a few short bursts of power before the bottle must be refilled—or supercharge or turbocharge. If power sustained at all engine speeds is the goal, supercharging or turbocharging is the only answer.

V-8 power mongers who just can't get enough normally-aspirated power can always install a turbo kit on their V-8, as this owner did to his Corvette. Note the cruise control.

Gale Banks has found that extreme turbo power necessitates drivetrain and suspension changes, hence his thoroughly engineered American Turbocar.

Both supercharging and turbocharging increase the mean effective cylinder pressure by pressurizing the air or fuel/air mixture in the intake manifold and cylinder. The obvious difference between these two pumping systems is the way in which they are driven. The turbocharger is driven by a turbine spun by expanding exhaust gases, while a supercharger is driven by a belt from the engine's crankshaft. This drivetrain difference in itself is an advantage of turbocharging: driving a supercharger soaks up engine horsepower, hurting fuel economy and placing more strain on the engine. Driving a turbocharger, on the other hand, wastes very little power, because it is spun by waste heat in the exhaust gases. A slight increase in exhaust backpressure wastes some engine power with the turbo, but because it is not only exhaust gas pressure but also exhaust gas temperature which drives the turbo, the power loss due to exhaust backpressure is small. In fact, at wider throttle openings, this pressure is more than compensated for by reduced pumping effort in inducting the incoming charge. The drop

The manufacturer of some of the most expensive turbocars, Mercedes-Benz, sold more turbos in America in 1983 than any other manufacturer—imported or domestic.

Garrett AiResearch is stepping up production of its automotive turbochargers to meet the demand for Ford, Chrysler, GM, and Mercedes turbocars.

in exhaust gas temperature of up to 300 degrees F. through the turbocharger represents the heat energy which drives the turbocharger. Without a turbocharger, the potential energy of this exhaust heat is wasted out the tailpipe into the outside air.

But the advantages of turbocharging over supercharging do not end there. Two types of superchargers are currently in popular use. The roots type, such as the hot rodder's popular GMC-based superchargers, hold an almost even boost pressure regardless of engine speed, but they only have a compressor efficiency of about 50%. As a result, they cause an increase in intake air temperature of roughly 40% more than a turbo. This means poor fuel economy, high engine heat, and a less-than-optimum power increase.

Every 1984 Mitsubishi model is available with a turbo.

Centrifugal superchargers, like the Paxton, have an even higher compressor efficiency than most turbochargers: about 80% efficiency. Unfortunately, their boost pressure rises roughly with the square of engine speed. So a centrifugal supercharger that puts out nine pounds of boost at 6000 rpm only puts out 3 pounds of boost at 3000 rpm—not very good for low and mid-range power. The only possible solution would be a variable-ratio drive—but that technology will have to wait till tomorrow. Centrifugal superchargers also weigh more than turbochargers and take up more space.

Turbochargers offer a compressor efficiency of around 70%, almost as good as a centrifugal supercharger. With a turbo, boost pressure also rises with engine speed, but nowhere near as much as with a centrifugal supercharger. The increase with a turbocharger is essentially linear, so that doubling rpm roughly doubles boost pressure. But because the turbocharger is driven by exhaust gases instead of a belt, there is a simple and efficient way to obtain a steady boost pressure regardless of engine speed. When the maximum safe boost level is reached, some exhaust is diverted from the turbine with a wastegate, reducing the compressor's speed just enough to hold a steady level of boost pressure. The diverted exhaust is dumped into the exhaust system beyond the turbocharger to travel through the muffler and out the tailpipe.

Some of the latest high-tech turbocars are taking this advantage one step further. The '84 Buick V-6 turbo, Pontiac 2000 turbo, Audi 5000S turbo, Ford Mustang SVO, and Saab turbo use electronically-controlled wastegates to constantly monitor and adjust boost pressure for optimum performance. This kind of control on a centrifugal supercharger would require a variable ratio drive combined with an electronically-controlled infinitely-variable pulley—a system we'll have to leave to the Leonardo DaVincis of the future.

So the big story in performance cars today is turbocharging, and nobody knows that fact better than the auto manufacturers, both foreign and domestic. A production supercharged car has not been available in America since the '60s, but there have been 51 turbocharged production cars available here since 1980, and 43 of those turbo models are available for '84.

495,000 turbocharged cars were sold in the United States between 1980 and 1983. The lion's share of these turbo cars, 279,000, were imports. While the import manufacturers sold only 20,000 turbos here in 1980, their sales were up to 118,000 turbos here in '83. The manufacturer of the most expensive turbo cars, Mercedes-Benz, led the way selling 42,700 turbo-diesels in '83, for a grand total of 112,000 between '80 and '83. Datsun has sold 36,625 280 ZX Turbos here, 16,308 of those in '83 alone. In late '83 Nissan added the Pulsar NX Turbo and the 280 ZX Turbo gets updated to a

Ford discontinued the Mustang/Capri turbo at the end of 1980, but has reengineered and reintroduced it for '84, along with a total line-up of nine Ford turbo models for '84.

The Buick Regal Turbo has been in production since '78, a record for domestic turbocars.

300 ZX Turbo for '84. 34,000 Saab 900 Turbos have been sold in the U.S. between '80 and '83, and 12,117 of those were sold in '83. 29,329 Volvo turbos were sold here in the '80–'83 period, and more than half of those turbos, 15,421, were sold in '83. Peugeot sold 4,575 turbos in the U.S. in '83, for a total of 22,353 turbo-diesels since 1980. Porsche has sold 8000 turbos in the U.S. since 1980, even though they currently have no turbo model available here. In their first-year effort, Mitsubishi sold 7644 turbos here in '83, and has added two more turbo models for '84, so that every Mitsubishi model is available as a turbo. Audi sold 7,150 turbos in '83, and is introducing their new 5000S Turbo for '84. Volkswagen sold 6106 turbo-diesels here in '83, and Renault sold 5662 Fuego Turbos in the U.S. in '83. And Subaru has just jumped on the turbo bandwagon with the 1800 4WD Turbo.

The domestic auto manufacturers sold 216,000 turbocharged cars between 1980 and 1983. But unlike the imports, domestic turbocar sales have been in decline since 1980. More than half of these domestic turbocars were sold in 1980 alone, as a number of manufacturers dropped out of the turbocar business after '80 or '81—at least temporarily. Ford, for instance, sold 42,000 turbo Mustangs and Capris in 1980, before discontinuing the turbo—only to reincarnate the turbo Mustang/Capri for '84. International Harvester, which built 10,381 turbo Scouts, Travellers, and Terras in 1980 permanently stopped production of this entire line at the end of '80. Pontiac built 23,421 turbo Trans Ams in 1980, and 16,487 in '81, before switching to less powerful nor-

Chrysler has never before offered a turbocharged car, but will have five turbo models beginning '84, including the all new Dodge Daytona Turbo.

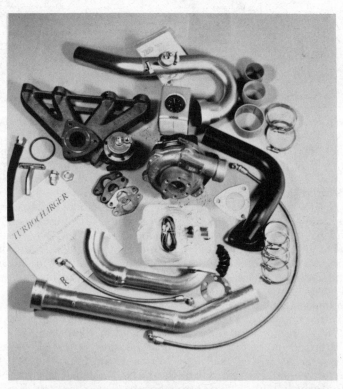

Eighteen turbo kit manufacturers are selling 180 different turbo kits, such as this Arkay kit designed to convert the gasoline VW Rabbit to turbo-operation.

mally-aspirated Chevy V-8s in the redesigned '82 Firebird. Chevy built 13,839 turbo Monte Carlos in '80, but production dropped to 1,951 Monte Carlo Turbos in '81, after which installation of the Buick V-6 Turbo in the Monte Carlo was stopped. The only steady domestic producer of turbocars has been Buick, and their turbo sales dropped from 37,000 in '82 to 15,700 in '83.

Despite these recent poor turbo sales, the domestic manufacturers are returning to the turbo fold in droves for '84. In fact, the domestic manufacturers' combined turbo-car sales projections for '84 are 278,000—more than the total domestic turbo sales for '80 through '83! The domestic manufacturers clearly see turbocharging as the answer to their performance option dreams. In Detroit, 1984 is the year of the turbo. Ford for instance, has nine models available with turbos for '84, compared with no turbo models since 1980. Ford expects to sell 125,000 turbo-cars in '84. Chrysler, which has never offered a turbo before, has seven turbo models available for '84, and also predicts total turbo sales of 125,000 for '84. Buick anticipates 16,000 turbo sales in '84, and Pontiac plans to sell 12,000 new 2000 Turbos.

But the factory turbo-car manufacturers are not the only ones banking on turbos. Eighteen aftermarket manufacturers currently market a total of 180 turbo kits which convert normally-aspirated cars into turbos, and a plethora of companies offer turbo accessories ranging from intercooler kits to boost gauges. So car enthusiasts rejoice: performance is back, stronger than ever, and we have that little turbine/compressor gadget to thank.

2
Turbo Technology

In a normally-aspirated engine the fuel/air mixture is sucked into the cylinder by the piston. As the piston goes down in its cylinder, it creates a vacuum which draws in the fuel/air mixture. The more fuel/air mixture the engine gets, the more power it will produce. But this normal aspiration will only allow a limited amount of fuel/air mixture into the cylinder. If the piston could suck in the cylinder's full displacement worth of fuel/air mixture with each piston downstroke, the engine would have 100% "volumetric efficiency." For example, if a 500 cubic centimeter cylinder could draw in the amount of fuel/air mixture that would fill a 500cc container at atmospheric pressure and temperature, that engine would have 100% volumetric efficiency.

But due to intake and exhaust system flow restrictions, most automobile engines can only muster around 80% volumetric efficiency. Some normally aspirated racing engines actually achieve slightly more than 100% volumetric efficiency over a narrow rpm band. Radical cam timing, big valves and carbs, and careful intake and exhaust port and manifold sizing can create a ram air flow effect, producing 100+% volumetric efficiency over a narrow range of engine speeds. But at higher or lower rpm, volumetric efficiency drops way down—and so does power. The price of such high volumetric efficiency with normal aspiration is a very narrow power band.

A turbocharged engine need not rely on the piston's vacuum to fill the cylinder with fuel/air mixture. Instead, the turbocharger forces the fuel/air mixture into the cylinder under pressure. In doing so, most turbocharged engines can exceed 120% volumetric efficiency over a wide rpm band. That high volumetric efficiency means the turbocharger is stuffing lots of fuel/air mixture into the cylinder, increasing mean effective pressure in the cylinder, and thereby producing more of the beloved horsepower. Of course, during low speed cruising, the turbocharger is not pressurizing the intake manifold, and the vacuum created by the piston's descent is still primarily responsible for engine breathing (though the turbocharger may reduce the effect of intake restriction).

There are five different intake systems commonly used on turbo gasoline engines. Every carbureted production turbocar, including the 79–80 Mustang/Capri, '78–82 Buick, and '80–81 Pontiac Trans Am, has used the draw-through carburetor system. In this design, the turbocharger sucks the fuel/air mixture through the carburetor (where the fuel and air are mixed) and then forces this mixture into the intake manifold. Roughly half the carbureted aftermarket turbo kits are draw-through, while the rest are blow-through carburetor designs. In this design, the turbocharger sucks in only air, and blows this air through the carburetor under pressure. There are pluses and minuses to each design, with no clear overall winner. With the blow-through system, the carburetor

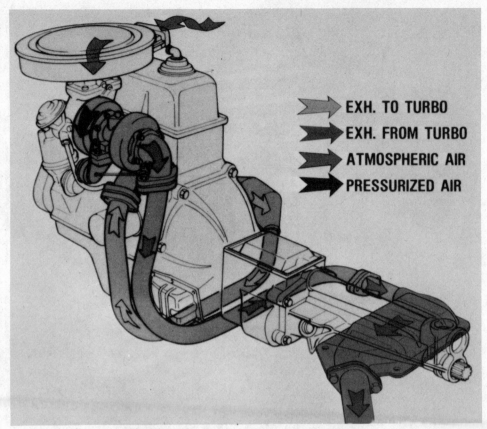

EXH. TO TURBO
EXH. FROM TURBO
ATMOSPHERIC AIR
PRESSURIZED AIR

In the draw-through carburetor design, the turbocharger sucks the fuel/air mixture through the carburetor and then forces the mixture through the intake manifold.

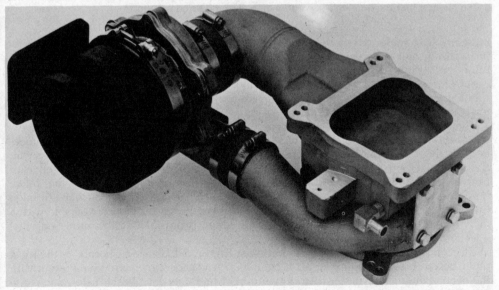

The draw-through design is easily understood by viewing the Spearco slope-flow module. The carburetor mounts on top of the module, and the base of the module bolts to the intake manifold.

A side-draft carburetor, as used in this Shankle Alfa-Romeo kit, allows the fuel/air mixture to be drawn straight into the turbocharger with a draw-through turbo system.

must not leak under pressure, the fuel pump pressure must rise as the carburetor is pressurized, and the vacuum source for vacuum assist brakes is unsteady. Also carburetor temperature varies with load, which affects the fuel/air mixture. Draw-through carburetor systems are more likely to suffer from carburetor ice-up, and unless a sidedraft carb is used (rare), the fuel/air mixture's route to the cylinders is slightly more tortuous.

There are currently no factory turbocars available in America with carburetors except the Maserati Biturbo. Virtually all the manufacturers have switched to fuel injection for their turbocars, and most aftermarket turbo kits for fuel injected cars retain the fuel injection. There are three basic fuel injection designs commonly used with turbochargers today: electronic port fuel injection, electronic throttle body fuel injection, and mechanical fuel injection. All of these designs are used with blow-through turbocharger systems. No factory fuel injected turbocar has ever been available with a draw-through turbo system, although such a design is possible with throttle body fuel injection.

Electronic port fuel injection, used in most late model factory turbocars, is the most sophisticated fuel injection system. It combines dead accurate electronic metering with the even cylinder to cylinder distribution of port fuel injection. Bosch L-Jetronic is the most widely used electronic port fuel injection system and many turbocar manufacturers such as Subaru, Pontiac and Chrysler, combine Bosch L-Jetronic parts with their

In the blow-through carburetor design, the turbocharger sucks air through the air filter and then forces it through the carburetor to the intake manifold.

own hardware and software. Other manufacturers, like Ford and Nissan, have started from scratch and developed their own systems. With any electronic port fuel injection system, air is first drawn through the air cleaner and air flow meter. The turbocharger then compresses the air and forces it through the throttle into the intake manifold. The throttle is a butterfly valve operated by the driver which controls the air flow. The injectors (one per cylinder) are mounted adjacent to each intake port in the intake manifold, and squirt the gasoline into the air just before it reaches the cylinders.

The electronic control unit tells the injectors how much fuel to inject, based on information from a number of sensors such as coolant temperature, engine speed, exhaust oxygen content, throttle position, and air flow. Most air flow sensors are trapdoor designs. Intake air pressure forces the trap door open and the trap door's position is translated into an electric signal which is sent to the control unit. At the same time, an intake air temperature sensor tells the control unit what the air temperature is, so the control unit can assess the air density. After all, what the control unit really wants to know is the weight of the intake air, not the volume because it's weight that determines how much fuel is needed. For this reason, and because the trap door restricts air flow, some manufacturers, like Buick with their '84 V-6 turbo, are switching to mass air flow sensors, which determine the weight of the air more simply and cause less intake restriction because they need no trap door.

Mitsubishi uses a flapless ultrasonic mass air flow sensor with their unique (for a factory turbocar) throttle body fuel injection. The ultrasonic sensor flows intake air past a triangular object that generates small eddy currents or vortices in the airflow with

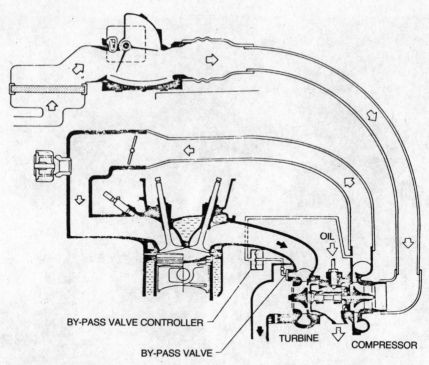

OIL

BY-PASS VALVE CONTROLLER

TURBINE   COMPRESSOR

BY-PASS VALVE

With electronic port fuel injection, air is drawn through the air filter and air flow meter to the turbocharger, which forces the air through the throttle to the intake manifold.

minimal restriction. Untrasonic (high frequency) waves are then generated electronically on one side of the air intake and counted on the opposite side to measure airflow.

Instead of using a separate injector for each cylinder, the Mitsubishi's throttle body has a pair of integral injectors which inject the fuel for all the cylinders inside the throttle body. So the turbocharger sucks air through the air filter/air flow sensor, compresses the air, and forces it through the throttle body where the butterfly controls air flow (via the driver's right foot) and the injectors squirt in the gasoline. The fuel/air mixture is then forced through the intake manifold and into the cylinders.

The only mechanical fuel injection system used in factory turbocars is the Bosch K-Jetronic system, and it is used by Volvo, Porsche, Audi, and Saab. Bosch K-Jetronic is a ported fuel injection system. Air is sucked through the air cleaner and air flow sensor. The air flow sensor is a trap door, and as air flows past the trap door, the door is forced open. The greater the volume of intake air, the more the trap door is opened. Unlike an electronic fuel injection system, the trap door position is not translated into an electric signal for an electronic control unit. Instead, a mechanical lever connected to the trap door varies the position of the control plunger in the fuel distributor, thus regulating fuel flow to the injectors. Intake air temperature is not measured (as it is in electronic fuel injection systems), so air density is not taken into account in adjusting the fuel/air mixture. The turbocharger compresses the air, and forces it through the throttle where air flow is controlled by a butterfly valve connected to the throttle pedal under the driver's right foot. Finally the air is piped to the intake manifold where injectors squirt fuel into the air just before the air enters the cylinders.

Most of the K-Jetronic systems use an oxygen sensor, which monitors the oxygen content in the exhaust. This information allows the fuel injection system to continuously fine tune the fuel/air mixture, and to some extent compensates for the K-Jetronic's lack of regard for intake air density. Most electronic fuel injection systems also use an oxygen sensor, and in these systems the sensor's electric signal is fed into the fuel injection system's main electronic control unit or microprocessor, which then alters its orders to the fuel injectors. But in the K-Jetronic system, the oxygen sensor signal goes to a minor-league electronic control unit which squirts additional fuel into the intake via a separate modulated injector when the engine is started and/or orders the fuel pressure regulator to alter fuel pressure to the main injectors, thereby changing the amount of fuel squirted into the intake air by the injectors.

The air intake system of any diesel engine is relatively simple, and a turbocharger only makes it one step more complicated. The diesel has no throttle, carburetor and usually no air-flow meter. Air is simply sucked through the air filter, compressed by the turbocharger, and forced through the intake manifold to the cylinders. The air doesn't meet the diesel fuel until the fuel is injected directly into the combustion chamber or prechamber. Some turbodiesel manufacturers, like Mercedes-Benz, use a boost pressure sensor in the intake manifold to assist the mechanical fuel governor in metering the fuel. This way a limit is placed on the amount of fuel injected, depending upon boost pressure. When boost pressure is very low, such as at low rpm, or for a period of a second or two after the driver opens the throttle, the fuel pump cannot inject the maximum amount of fuel, but only that amount that can be burned cleanly with the amount of air actually available.

There are five brands of turbochargers commonly found on turbodiesels and turbo gasoline engines today. The Garrett AiResearch turbochargers, which feature integral wastegates, have been by far the most popular turbos for original equipment manufacturers, but their marketing in the aftermarket is almost nonexistent. The most common AiResearch turbo is the mid-sized T3 found in Buick, Mercedes, Volvo, Saab, Datsun, Ford, Chrysler, Pontiac, Chevrolet, Renault, Peugeot, and International Harvester turbo vehicles. The smaller T2 is now being used in the Pontiac 2000 and Nissan Pulsar turbos, and Nissan is using the big T5 in their powerful 300ZX Turbo.

The IHI turbochargers are becoming increasingly popular for both original equipment and aftermarket turbo kit use because of their compact, lightweight design, low rotating inertia, and efficient, backward-curved compressor wheels. They are particularly well-received on small engine applications, such as Subarus and Ford's 1.6 liter turbo. The IHI turbocharger is extremely popular in the aftermarket, thanks to its integral wastegate, a feature which the other popular aftermarket turbocharger, the Rajay, does not have. Rajay uses a die-cast aluminum bearing housing to keep turbo-

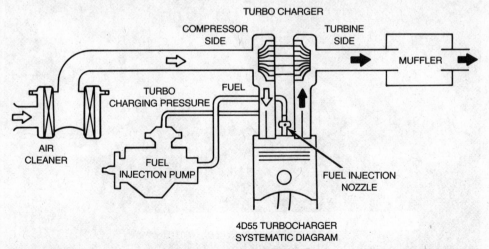

In a turbo-diesel, air is sucked through the air filter, compressed by the turbocharger, and forced through the intake manifold. Diesel fuel is injected into the combustion chamber or prechamber. There is no throttle.

TURBOCHARGER

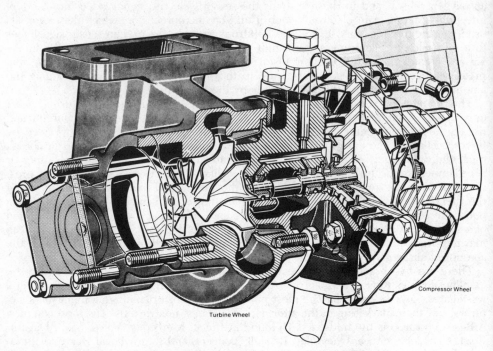

Turbine Wheel

Compressor Wheel

The Garrett AiResearch turbocharger is common on factory turbocars.

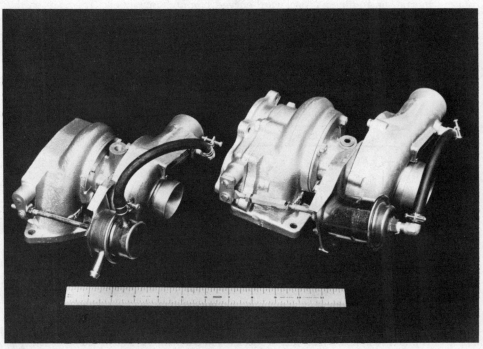

IHI turbochargers are compact.

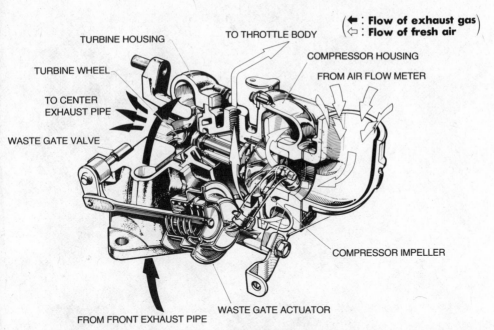

$\left(\begin{array}{l}\text{←} : \text{Flow of exhaust gas}\\ \text{⇦} : \text{Flow of fresh air}\end{array}\right)$

TURBINE HOUSING

TURBINE WHEEL

TO CENTER
EXHAUST PIPE

WASTE GATE VALVE

TO THROTTLE BODY

COMPRESSOR HOUSING

FROM AIR FLOW METER

COMPRESSOR IMPELLER

WASTE GATE ACTUATOR

FROM FRONT EXHAUST PIPE

The IHI's integral wastegate has made it the new darling of OEM and aftermarket manufacturers.

charger length to a minimum. Though still common in aftermarket kits, the Rajay has been falling out of favor recently as kit manufacturers switch to the high tech IHI. Roto-Master turbochargers are used by some kit manufacturers, and these are very similar to the Rajays, and also do not have integral wastegates.

Porsche and Audi use KKK turbochargers, which don't have integral wastegates, but do feature the unique Porsche-patented pop-off valve, which improves throttle response by eliminating compressor stall when the throttle is closed while under boost. Instead of air deadheading against the throttle when the throttle is shut, this device routes the air from the compressor discharge back to the compressor inlet, keeping the compressor spinning. When the throttle is reopened, the compressor is still spinning fast, so it needs less time to accelerate up to the speed needed to build boost. DM

The tried and true Rajay turbocharger is popular in the aftermarket.

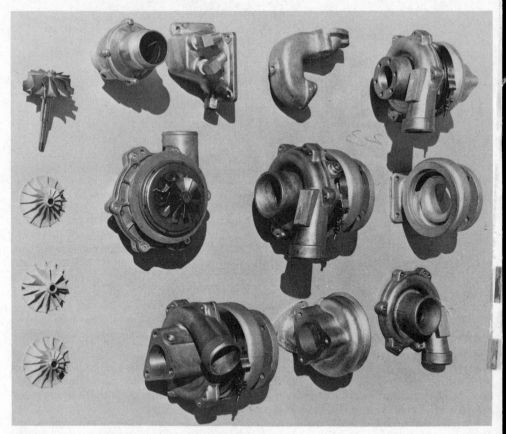

Rajays are available for many different applications.

Rotomaster turbochargers are used in some aftermarket kits. They are similar to Rajays.

Porsche and Audi use KKK turbochargers which feature a pop-off valve to improve throttle response.

Engineering's Dayton valve, which mounts to any turbocharger for blow-through throttle fuel injected systems, is designed to achieve a similar effect, and Blake Enterprises manufactures a compressor bypass kit for short inlet Rajay turbochargers.

Mitsubishi manufactures its own turbochargers, and currently has four models for automotive use: the TC03, TC04 (used in the Dodge Colt Turbo), TC05 (used in the Mitsubishi Tredia, Cordia, and Starion turbos), and TC06. These model numbers correspond roughly to the diameter of the turbine wheel in centimeters and the smallest of these, the TC03, will fit comfortably in the palm of your hand. It is also the world's smallest turbocharger to date and the turbine wheel and compressor are designed to tolerate rotational speeds of up to 210,000 rpm. Mitsubishi's turbine wheels and compressors are precision cast by the lost-wax method and then attached to the shaft that joins them by electronic beam welding. Each unit is hand finished to a tolerance of one one-thousandth of a millimeter and balanced by computer.

Any turbocharger can be divided into three general sections: turbine, driveshaft, and compressor. The turbine converts exhaust energy into rotational motion, and the driveshaft transmits this motion to the compressor, which compresses the air or fuel/air mixture. Exhaust gases enter the exhaust gas inlet of the turbine housing and expand, picking up speed, spin the turbine wheel, bound off the exducer portion of the turbine wheel, and exit the turbine via the exhaust gas outlet.

An important consideration for engineers who must match a turbocharger to a particular engine is the A/R ratio, which affects the turbine speed for a given exhaust flow. The A/R ratio is the ratio between the turbine nozzle area and the distance from the

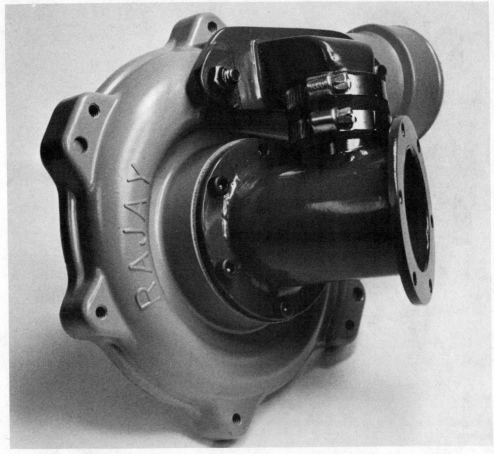

Blake Enterprises' compressor bypass keeps the Rajay's compressor spinning when the throttle is shut.

center of the turbine wheel to the centroid (or center point) of the nozzle area. A larger A/R ratio causes the turbine to spin slower. This ratio can also be useful to a car owner who is contemplating modifications. If his current A/R ratio is say .7, and he wants to increase turbine speed, he can use a turbine housing with an A/R ratio of .6. Most turbine housings have their A/R ratio stamped or cast in for easy identification, and switching turbine housings is an easy procedure on engines with easily accessible turbochargers.

The turbine wheel spins the driveshaft, which sits inside the center bearing housing. There are aluminum or bronze bearings in the center housing bore and an oil seal at each end. The bearings are not pressed into the housing, because such a rigid grip causes severe bearing and journal wear due to high speed turbine shake. Instead, floating or semi-floating bearings are used. Oil enters the oil inlet under pressure and flows through and around the bearings. Full film lubrication produces a complete physical separation of the spinning journal and bearing surfaces. Oil flowing between the bearing and bearing housing damps out vibration caused by minute turbine wheel imbalances. Oil also reaches the thrust bearing surfaces before flowing out the oil drain fitting.

The driveshaft spins the compressor wheel which sucks the fresh air or fuel/air mixture into the turbocharger with its inducer, the gas entry portion of the compressor

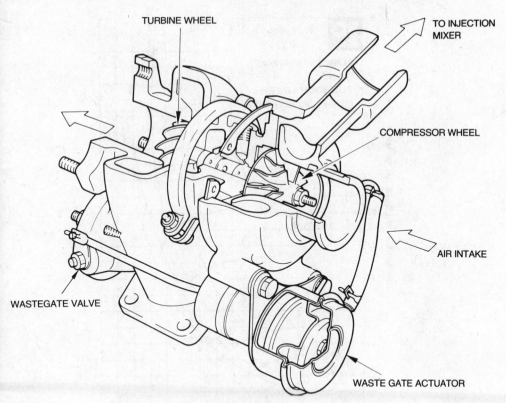

TURBINE WHEEL

TO INJECTION MIXER

COMPRESSOR WHEEL

AIR INTAKE

WASTEGATE VALVE

WASTE GATE ACTUATOR

Mitsubishi turbochargers are compact and have integral wastegates.

wheel. The compressor wheel and housing then work together accelerating and compressing the air or fuel/air mixture and forcing it downstream.

Compressors can be designed to give high flow over a narrow rpm range or moderate flow over a broad rpm range. A narrow range compressor is very efficient in its narrow flow range, but inefficient at speeds above and below this range. This inefficiency causes high intake air temperatures, lowering air density and thus power. By the same token, a broad range compressor is only moderately efficient at any speed. Thus some compromise between these two extremes is the best solution, and the most common compromise is a flow range of around 2:1. This means that at a 2:1 compressor pressure ratio, maximum flow (calculated at 60% compressor efficiency) is twice the flow obtainable at the surge limit. The surge limit is the minimum stable flow, below which compressor output is unsteady. The compressor pressure ratio, which must not be confused with the flow range, is the ratio between compressor output pressure and compressor intake pressure. This pressure ratio increases with compressor speed. Flow range is much easier to visualize when using a compressor map, which is a chart showing a specific compressor's flow characteristics.

Because a turbocharger big enough to give the desired maximum flow on a V-8 engine is generally too big to give optimum throttle response, V-8s are frequently fitted with two turbochargers. Occasionally six cylinder engines also receive the twin turbo treatment. With twin turbos, smaller turbochargers can be used, and their reduced inertia and more sensitive flow characteristics improve throttle response on a big engine in comparison with a single turbo system. Intake air temperature tends to be lower with a twin turbo set-up at the same flow level because of improved efficiency. Another advantage of using twin turbos on V-8s and V-6s is elimination of an exhaust

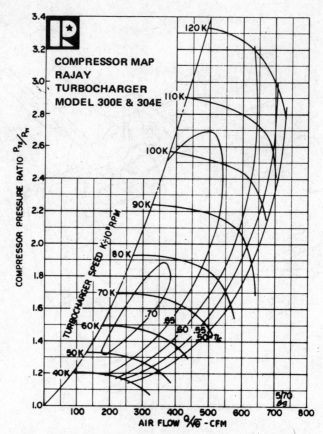

**COMPRESSOR MAP
RAJAY
TURBOCHARGER
MODEL 300E & 304E**

Compressor maps allow the turbo engineer to visualize a compressor's flow characteristics.

crossover pipe, shortening and equalizing the exhaust gases' trek to the turbocharger. Dual exhausts can be used with twin turbos for improved exhaust flow.

Detonation is the prime power limitation in turbocharged gasoline engines. Detonation (also known as "ping" or "knock") is the uncontrolled spontaneous explosion of fuel/air mixture in the combustion chamber. Normally, the fuel/air mixture burns quickly in the combustion chamber, raising the fuel/air mixture's temperature and hence pressure, forcing the piston down in the cylinder. But when detonation occurs, part of the fuel/air mixture explodes, causing a loss of power. Instead of a smooth flame front advancing from the spark plug, detonation causes a violent collision of flame fronts. Multiple flame fronts result from ignition occuring not only as a result of spark plug action, but that of prolonged high temperatures in other areas of the combustion chamber causing spontaneous combustion. No engine can survive sustained detonation, as the physical and thermal stresses inflict damage to the piston, sparkplug, rod, and bearings. In many cases, blown head gaskets can be traced to detonation.

Sustained detonation produces a severe rise in temperature, with resultant weakening and removal of metal from the piston crown. Severe detonation is most likely to hammer a hole through the piston. The hole is usually sharp-edged. Radial cracks and a depressed area may be found adjacent to the actual break, and evidence of excessive temperature can sometimes be seen in pitting of metal on the top surface of the piston. Broken ring lands or cracks in the piston wall may be in evidence, and broken or split spark plug firing ends are usually indicative of detonation.

Twin turbos improve throttle response on big engines.

Detonation can lead to preignition. When preignition occurs, a spark is not required to ignite the fuel. A part within the combustion chamber has reached the point of incandescence, or the fuel (due to its instability under pressure and heat) simply ignites without spark. The difference is that the spontaneous combustion begins before the spark plug fires instead of somewhat after.

The extreme temperatures associated with preignition generally result in melting the piston. The edges of the breakthrough indicate a typical thermal failure unlike the mechanical failure caused by detonation. Spark plugs exposed to sustained preignition

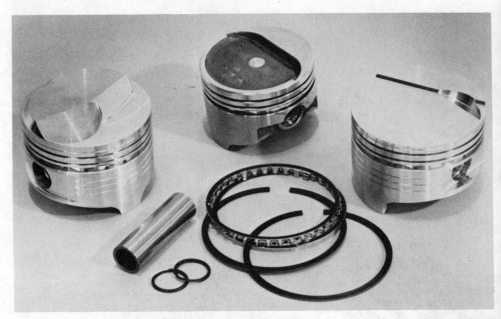

Low compression forged pistons, such as these Gale Banks Chevy small block pistons, reduce the chance of engine damage caused by detonation.

temperatures are likely to have their center electrodes melted away. In the extreme cases, the ceramic insulator tip may appear fused.

Detonation is caused by a number of factors, including high compression ratio, high boost pressure, low gasoline octane, advanced ignition timing, lean fuel/air mixture, high fuel/air mixture temperature, or improper combustion chamber/piston crown design, or camshaft timing. Some of these factors, such as compression ratio, combustion chamber/piston design and camshaft timing, are designed into an engine, and are not easily adjustable or controllable. The compression ratio is the ratio between the volume in the cylinder when the piston is at the bottom of its stroke and the volume in the cylinder when the piston is at the top of its stroke. A higher compression ratio produces more power from a given amount of fuel/air mixture, but the limit is detonation. Because a turbo forces more fuel/air mixture into the combustion chamber while under boost pressure and the fuel/air mixture is at a higher temperature, a turbocharged engine must use a lower compression ratio than a comparable normally-aspirated engine. While the average compression ratio for a normally aspirated gasoline car engine is 8.7:1, for a turbo it is 7.6:1. Overall efficiency does not suffer with the turbo partly because the air/fuel is actually compressed as much in a turbocharged engine as in a normally aspirated one. Though the numerical compression ratio, representing what occurs in the cylinder, may be lower, the higher initial pressure may mean the mixture is actually compressed as much or even more than in the standard engine.

Combustion chamber/piston crown design is usually pretty well optimized by the manufacturer, taking into account valve and sparkplug arrangement, swirl and flow patterns, etc. Camshaft timing is important because the intake valve stays open during the early stages of the compression stroke. Therefore, the effective compression ratio is as dependent on the intake valve timing as it is on the compression ratio itself. The longer the intake valve stays open, the lower the effective compression ratio. So retarding the camshaft (or designing a new camshaft with a retarded intake event) has the twofold effect of placing the engine's powerband at a higher rpm and reducing detonation.

Some of the factors that contribute to detonation are controllable. A cooler fuel/air

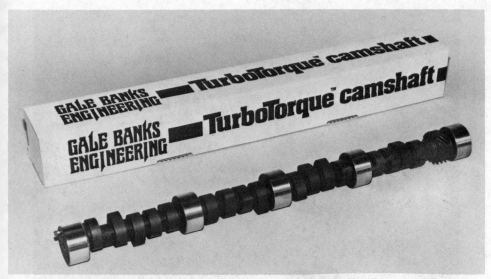

Camshaft timing affects detonation, and a camshaft that is properly matched to a turbocharged engine will yield optimum performance.

mixture is less likely to detonate than a hotter mixture. Further, a cooler mixture is a denser mixture, which means more power-producing fuel/air mixture in a given volume. Fuel/air mixture temperature is particularly critical in a turbocharged engine, because the turbocharger increases the air or fuel/air mixture temperature as a by-product of the compression process (Boyles law). A turbocharger will typically increase the air or fuel/air mixture temperature by 150°F or more under boost conditions. This means that on a 75°F day, the air or fuel/air mixture is 225°F before it reaches the cylinder head. An intercooler can typically reduce this temperature by 90°F, bringing intake air temperature down to say 135°F.

An intercooler is placed between the turbocharger's compressor discharge and the cylinder head, and looks like a small radiator. Two types of intercoolers are commonly used: air-to-air and air-to-water. In an air-to-air system, the compressed air or fuel/air mixture is forced through the intercooler's core, and cool ambient air flows through the fins, cooling the compressed air or fuel/air mixture. With air-to-water systems, the compressed air is forced through the intercooler's fins, and antifreeze is piped through the cores. Air-to-water systems vary somewhat, in that some integrate the intercooler system into the engine coolant system by using the engine coolant, engine water pump, and engine coolant radiator, while others completely segregate the intercooler system from the engine coolant system by using a separate (usually electric and boost pressure-activated) waterpump, coolant, and mini-radiator.

Positioning of an air-to-air intercooler or air-to-water radiator is critical for maximum heat transfer. Popular locations include: beneath a hood scoop or louvers, inside a fender, and in front of the engine coolant radiator. Some designs also use an electric fan to increase air flow through the fins at low speeds.

An additional payoff of an intercooler is reduced exhaust gas temperatures, which makes life easier on the exhaust valves. In fact, every one degree reduction in intake air temperature is good for roughly a one degree reduction in exhaust gas temperature, a phenomenon which occasionally requires a slightly smaller turbine housing to keep turbocharger speed up. There is also a loss in boost pressure through the intercooler, usually about one psi. However, this pressure loss is much more than made up for powerwise with the increased intake air density.

Boost pressure, the pressure in the intake manifold, is controllable, and the most

Callaway's Volkswagen 1.8 liter kit features an air-to-air intercooler.

Spearco offers an air-to-water intercooler for their DOHC Toyota Supra kit.

AK Miller has a selection of air-to-air intercoolers using AiResearch cores and O-ring sealed swivel connectors.

popular method of control is the wastegate. More boost pressure means more fuel/air mixture in the cylinders, and hence more power. As we said, the limit is detonation. Turbochargers put out more boost pressure at high rpm than at low rpm. A turbocharger sized to never exceed the maximum safe boost pressure at high rpm would put out little or no boost pressure at low rpm. Consequently, turbochargers are virtually always sized to give good boost pressures at low rpm, and a wastegate is used to bleed off excess exhaust gases, limiting peak turbocharger rpm, and limiting the maximum boost pressure. In the typical wastegate system, a pressure hose leads from the intake manifold to the wastegate diaphragm. When intake manifold pressure exceeds the preset limit, manifold pressure pushes the wastegate diaphragm against the spring, opening the wastegate valve to let exhaust gases bypass the turbine. When manifold pressure goes down, the spring recloses the wastegate valve. Adjustable spring preload adjusters, commonly called dial-a-boosts, are available for some popular wastegates, allowing the owner to adjust boost pressure manually.

Fuel octane is a critical factor in the elimination of detonation. Higher octane gasolines resist detonation. The most cost-effective fuel on the market today is the common 89 octane leaded regular. It combines low cost with a decent octane rating. Unfortunately, this fuel clogs and ruins catalytic converters. So most late model car owners in the U.S. must choose between expensive, low 87 octane unleaded regular, and super expensive 90 to 93 octane premium unleaded. Many factory turbocars are designed to run on premium unleaded, but will accept unleaded regular with reduced power as a consequence. Mixing unleaded premium 50/50 with leaded regular (also a no-no with a catalytic converter) causes an octane jump of roughly one octane point over the premium's rating. But who wants to play musical pumps at a crowded station over a lousy octane point? Airport fuel stops (for 98 octane aircraft gasoline) are impractical, especially on a trip, and airport vendors generally don't like to sell to street cars, because they can get in trouble for not charging road use taxes. Buying drums of racing gasoline is fine for local cruising, but impractical on a trip. Octane boosters, which can be used on a trip, will raise the octane of unleaded premiums another two or three points (for a total of 92 to 95 octane), but you'll end up spending nearly two dollars per gallon. All in all, the alternatives to regular gasoline are pretty poor.

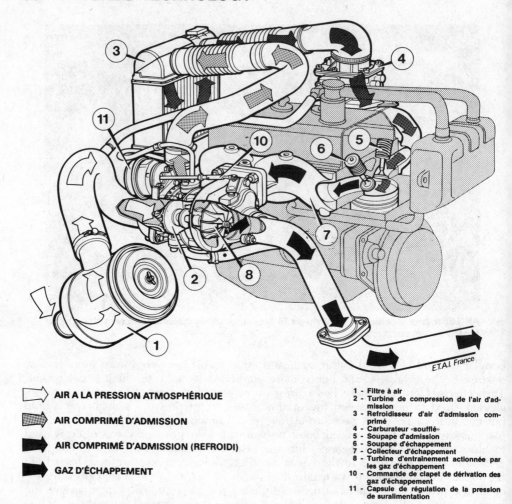

AIR A LA PRESSION ATMOSPHÉRIQUE

AIR COMPRIMÉ D'ADMISSION

AIR COMPRIMÉ D'ADMISSION (REFROIDI)

GAZ D'ÉCHAPPEMENT

E.T.A.I. France

1 - Filtre à air
2 - Turbine de compression de l'air d'admission
3 - Refroidisseur d'air d'admission comprimé
4 - Carburateur «soufflé»
5 - Soupape d'admission
6 - Soupape d'échappement
7 - Collecteur d'échappement
8 - Turbine d'entraînement actionnée par les gaz d'échappement
10 - Commande de clapet de dérivation des gaz d'échappement
11 - Capsule de régulation de la pression de suralimentation

Renault's inexpensive Fuego Turbo has an intercooler.

Water injection will cure the detonation blues (even with 87 octane unleaded gas), but the systems are temperamental, and adjusting them to inject the right amount of water at the right time is a tricky art. The water container must be refilled at each gas stop, and alcohol added in freezing weather. Also, you should use chrome or stainless steel valves with water injection to eliminate the chance of valve corrosion.

Enriching the fuel/air mixture, so that a greater proportion of fuel is mixed with the air, will reduce detonation up to a point. Go too rich, and you're just wasting fuel without any further detonation reduction. A number of aftermarket manufacturers use fuel enrichment systems in their turbo kits for fuel injected cars to reduce detonation as well as to provide adequate fuel under boost. Callaway can set up their Microfueler fuel enrichment system for any application. Holley makes a driver-adjustable fuel enrichment system for the electronically-controlled Holley-Weber two-barrel carburetors.

Advancing the ignition timing, so that the sparkplug sparks earlier during the piston's upstroke, improves power and fuel efficiency up to a point. Unfortunately, an advanced ignition timing is more likely to detonate. An engine can tolerate more ad-

Ford's Mustang SVO intercooler: the first intercooler on an American production car.

Audi's 5000S Turbo intercooler.

Volvo's optional intercooler kit is available from any Volvo dealer.

Rajay wastegate mounted to a three-bolt housing adapter available through Turbo City.

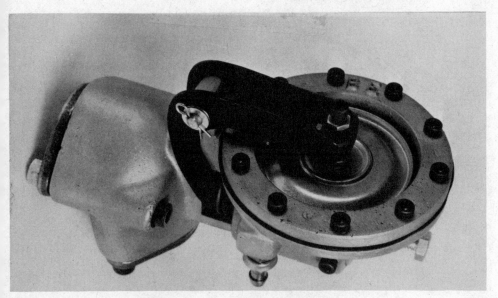

Blake's Boost Pressure Modulator is an adjustable wastegate.

TC-2 controller limits boost pressure but increases intake temperature.

Dial-a-Boost, available through Turbo Tom's, lets the driver quickly adjust maximum boost pressure. It works with Rajay and AiResearch wastegates.

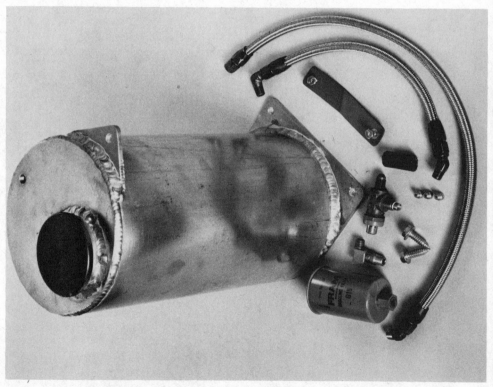

Turbo Tom's water injection system also lets boost pressure do the work. Holley jet meters the water. Tank is aluminum.

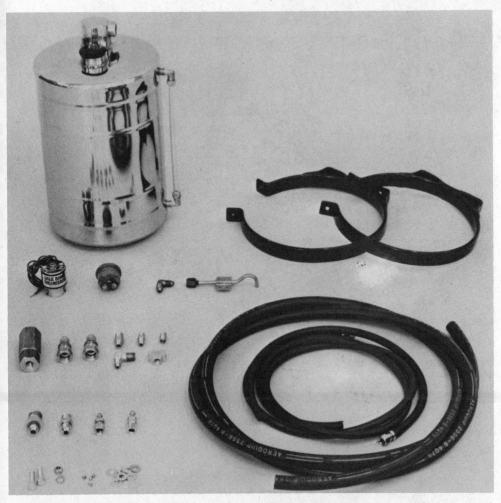

Gale Banks water injection system uses no pump. Boost pressure does the work. Stainless steel tank holds three gallons.

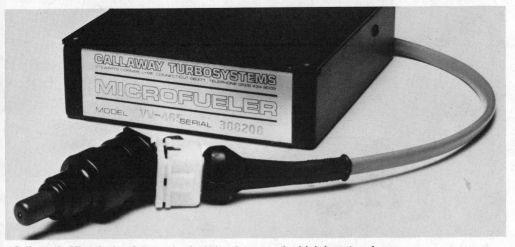

Callaway's Microfueler richens the fuel/air mixture under high boost and rpm.

Arkay's Detonation Control injects water and retards ignition timing in steps as boost pressure increases.

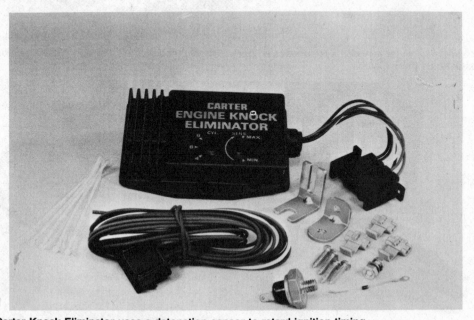

Carter Knock Eliminator uses a detonation sensor to retard ignition timing.

vanced timing while not under boost. Consequently, most turbos use some form of variable ignition timing system. Many aftermarket turbo kits use a boost-activated ignition retard system, which retards the timing while under boost. Some of these systems, like the Spearco and Arkay, are also available separately, and retard the timing in steps, dependant on boost pressure and/or engine speed. Most factory turbocars use a detonation-activated ignition retard system, in which a detonation sensor detects the frequencies associated with detonation and usually signals a computer to retard the timing. These systems vary from model to model, and are covered in detail in Chapter III. The most sophisticated electronic control systems, found on such turbocars as the Ford Mustang SVO, Pontiac 2000, Audi 5000S, '84 Buick Regal and Riviera, monitor engine functions with a dozen or more sensors, and use computers to retard ignition timing, reduce boost pressure, and adjust the fuel/air mixture as needed. These sophisticated systems provide maximum power and fuel economy with a minimum of detonation.

3
Turbo Gasoline Cars—
Factory-Direct

In 1980, there were nine turbocharged gasoline production cars for sale in America. As this is written, there are 37 turbo gasoline production cars, with new turbo models coming out nearly every month. Every turbocar imaginable from two-seater to station wagon is now for sale. This chapter reviews all of them and points out each model's pluses and minuses. Each car's turbo technicalities and innovations are covered, as are the hot options. So salivate over your favorites, boo the underachievers, and pick up some turbo facts.

The Maserati Biturbo uses a small IHI turbocharger mounted right near the exhaust manifold on either side of its V6 engine. This permits short exhaust runs for high turbocharger efficiency. Two smaller turbos respond much quicker than one large one, too. A single Weber carburetor is mounted inside the plenum chamber visible at the top/center of the photo.

AUDI 5000T/5000S TURBO

When Audi turbocharged the 5000 in 1980, 0 to 60 mile per hour acceleration was quickened by two and a half seconds, top speed went up by eight miles per hour, and EPA city gas mileage went up by one mile per gallon. The five cylinder gasoline engine's horsepower was boosted to 130 horsepower, thanks to the turbo, up from the normally aspirated engine's 103 horsepower. The engine's compression ratio was dropped from 8.0:1 to a low 7.0:1 for the turbo. Sodium-filled exhaust valves, an oil cooler, and five jets which spray more than a gallon of oil per minute against the undersides of the pistons keep the turbo engine cool.

Spindles and transaxle half shafts were toughened for turbo duty, and the T also got bigger brakes front and rear, alloy wheels, and Fulda 205/60 HR x 15 tires. The axle ratio was dropped from 3.90:1 to 3.73:1 for the turbo, slightly reducing the Audi's engine speed while cruising.

For 1984 the Audi 5000 Turbo has been vastly updated. The most obvious change with the new 5000S is the radically curved sheetmetal. The difference in appearance is as shocking as Ford's Fairmont-to-Tempo overnight swap. The aerodynamic drag improvement has reduced the amount of horsepower required to keep the Audi moving at 50 mph from 15.5 with the 5000T to 14.0 with the new 5000S. The 5000S is slightly longer and wider than the old T, and the S has seven cubic feet more interior room, and more trunk space, too.

The 5000S Turbo has 140 horsepower, ten more than its predecessor. A higher compression ratio (up from 7.0:1 to 8.2:1) an air-to-air intercooler, and electronic controls are responsible for the power increase. The Audi 5000S Turbo is the only mass-marketed imported car with electronically-controlled ignition timing and boost pressure. Maximum boost pressure is actually down to 7.2 psi from the old T's 8.7 psi.

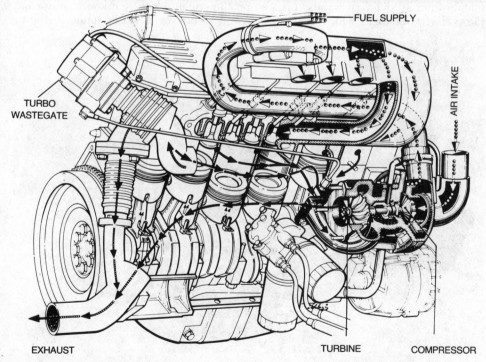

FUEL SUPPLY

TURBO WASTEGATE

AIR INTAKE

EXHAUST   TURBINE   COMPRESSOR

Five cylinder Audi 5000T engine has a low 7.0:1 compression ratio, sodium filled exhaust valves, and oil spray jets under each piston.

Audi's '84 5000S features curved sheetmetal for improved aerodynamics. Top speed is up 15 mph to 126 mph.

Audi has elected to take advantage of the electronically-controlled wastegate by increasing the compression ratio, rather than the more common practice of upping the maximum boost pressure. This approach yields maximum benefits in the area of fuel economy.

The electronic control unit, a three-chip micro-processor, controls the turbo wastegate control valve, ignition timing, fuel pump, oxygen sensor frequency valve, cold-start valve, digital–boost gauge, and the tachometer. The control unit processes information from 10 different inputs. Located on the engine block, the knock sensor has a piezo-electric element which generates current whenever it senses sound in the nar-

5000S intercooler sits in front of the engine, reducing the compressed intake air's temperature by 90° for 10 extra horsepower over the old 5000T.

1984 Audi 5000S Turbo Electronic Engine Control

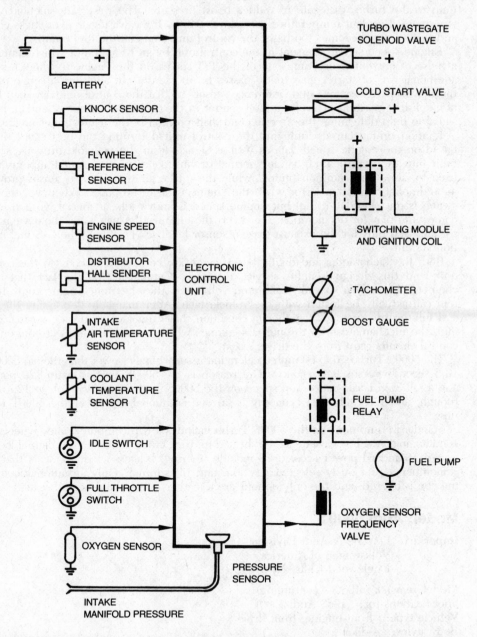

The 5000S is the only mass-marketed imported car which retards ignition and reduces boost when detonation is present. Supplied with information from the sensors on the left, the ECU controls all the functions on the right.

row frequency range associated with potentially damaging detonation. This system allows the engine to operate safely, and more efficiently, right at the threshold of detonation.

Whenever detonation is detected, ignition timing is retarded 2.7 degrees. If detonation stops, timing is advanced back in steps of 1.3 degrees, but if it continues, timing can be retarded in two more stages of 2.7 degrees. If detonation still continues with the full 8 degrees retard, the control unit then energizes a solenoid which applies vacuum to the turbo wastegate to reduce boost pressure. The wastegate solenoid valve also opens if coolant temperature goes above 246°F. If the wastegate ever sticks closed and turbo boost becomes too high, the control unit shuts off the fuel pump.

Engine speed signals are sent to the control unit by an RPM sensor which generates a pulse of current every time one of the 135 spikes on the engine flywheel (like an additional row of starter gear teeth) passes by the sensor. In addition, a single pin in the flywheel triggers a second "reference sensor" so that the control unit can keep track of crankshaft position. Thirdly, a Hall sender in the ignition distributor sends a phase signal to help determine the precise crankshaft position of the five-cylinder engine.

Ignition timing curves built into the control unit determine the best spark timing based on the engine speed data as well as signals from the manifold pressure sensor built into the control unit. Under normal driving conditions, coolant temperature is used to "fine-tune" ignition timing, while the intake air temperature sensor provides an additional correction factor when the engine is on turbo boost. Whenever the idle switch is on (closed throttle) but engine speed is above idle, a special ignition curve comes into play for deceleration. However, the full-throttle switch and the oxygen sensor are used only for the exhaust oxygen sensor feedback portion of the CIS fuel injection system.

Both the tachometer and digital boost gauge are connected directly to the control unit, and this design benefits service diagnosis. Capitalizing on the fact that a tachometer is really a specially-calibrated voltmeter, the electronic engine control system can actually be tested with the vehicle tach. After manually depressing the idle and full-throttle switches together (a condition that cannot occur during normal driving), the mechanic can go through a sequence which checks all the electronic engine control circuits right from the driver's seat.

The 5000S Turbo's 0 to 60 mph acceleration is surprisingly slower than the old 5000T's: 10.5 seconds versus 9.4 seconds. The reason is the 5000S Turbo's extra 120 pounds and a tall 3.25:1 axle ratio. Top speed on the 5000S Turbo is up 15 mph to 126 mph, though, and EPA city fuel economy is up one additional mpg over the 5000T to 19 mpg.

Standard equipment on the 5000S Turbo includes a trip computer, alloy wheels, air conditioning, AM/FM cassette stereo, heated mirrors, cruise control, and a central locking system. Standard power accessories include steering, brakes, windows, seats that will remember a previously selected position, and tilt sunroof. Only an automatic transmission is offered, and the only options are a leather interior and heated seats.

Model: Audi 5000T

Importer: Porsche & Audi Division
Volkswagen of America
Englewood Cliffs, NJ 07632

Model year (s):   1980—(continuing)
Specifications for: 1983 Audi 5000T
Vehicle type: front engine, front drive
Body styles: 4-door sedan

Wheelbase: 105.5 in
Length: 188.9 in
Width: 69.6 in

Height: 54.7 in
Weight: 2980 lbs
SAE volume; interior/trunk: 90/15
Fuel tank capacity: 19.8 gal

Engine type: 5 in-line turbo gasoline
Displacement: 2140cc/130.6 cu in
Turbo: KKK with separate wastegate
Maximum boost: 8.7 psi
Compression ratio: 7.0:1
Fuel system: Bosch K-Jetronic mechanical fuel injection
Horsepower: 130 @ 5400 rpm

Transmission (s):   3-speed auto

Front suspension: ind, MacPherson strut, coil springs
Rear suspension: rigid axle, coil springs

Acceleration, 0–60 mph: 9.4 sec
Top speed: 111 mph
Roadholding: .77 g
Braking, 70–0 mph: 214 ft
EPA Fuel economy, city: 18

Model: Audi 5000S Turbo

Importer: Porsche & Audi Division
Volkswagen of America
Englewood Cliffs, NJ 07632

Model year (s):   1984—(continuing)
Specifications for: 1984 5000S Turbo
Vehicle type: front engine, front drive
Body styles: 4-door sedan

Wheelbase: 105.8 in
Length: 192.7 in
Width: 71.4 in
Height: 54.7 in
Weight: 3100 lbs
SAE volume, interior/trunk: 97/17
Fuel tank capacity: 21.1 gal

Engine type: 5 in-line turbo gasoline
Displacement: 2144cc/130.8 cu in
Turbo: KKK with electronically-controlled wastegate, intercooler
Maximum boost: 7.2 psi
Compression ratio: 8.2:1
Fuel system: Bosch K-Jetronic mechanical fuel injection
Horsepower: 140 @ 5500 rpm

Transmission (s):   3-speed auto

Front suspension: ind, MacPherson strut, coil springs
Rear suspension: rigid axle, coil springs

Acceleration, 0–60 mph: 10.5 sec
Top speed: 126 mph
Roadholding: .75g
Braking, 70–0 mph: 214 ft
EPA Fuel economy, city/hwy: 19/28

AUDI QUATTRO

The Audi Quattro is unique. It's the only 4 wheel drive sedan with an intercooler. Except for the AMC Eagle, it's the only sedan with full time 4 wheel drive. It's the only 4 wheel drive sedan with a manual control for locking the center and rear differentials. And it's the quickest 4 wheel drive you can buy—even faster than a 454 big-block powered 4 wheel drive Chevy Suburban.

For all its uniqueness, the Quattro has strengths and weaknesses like most cars. Its strengths include 7.8 second 0 to 60 mph acceleration, a 122 mph top speed, and excellent brakes. Its snowy road performance is only average for a 4 wheel drive with the stock tires, but a switch to snow tires makes it excellent in the snow. On the negative side, the Quattro's 17 mpg EPA city fuel economy is too much like that of a V-8, its luggage room is too much like that of an RX-7, its roadholding around a smooth turn is mediocre, and its price is too much like that of a Mercedes.

The Quattro's engine is the Audi 2144cc five cylinder. An air—to—air intercooler mounted below the right side of the engine and 12.3 psi of boost bring horsepower up to 160 on the Quattro, though. Premium gasoline is required in the Quattro, and a special Hitachi ignition system adjusts the ignition timing as a function of engine speed, coolant temperature, engine load, etc. Titanium valve guides improve heat dissipation, and a larger oil cooler is used.

While the 5000T was only available with an automatic transmission, the Quattro is only available with a 5-speed manual. A center differential, which splits the power front and rear, is located at the rear of the transmission. The rear suspension is based on the 5000T's front suspension, using the same half-shafts and control arms and similar struts and spindles. The points which act as engine mounts up front serve as differential mounts in the rear. Four wheel disc brakes are used. A single hydraulic pump provides the power for brakes and steering. A knob on the dash console lets the driver lock the center and rear differentials even at speed, improving straightline traction for acceleration and braking. Cornering ability is best in the unlocked mode. The front differential is not lockable, as this would make steering difficult.

The quickest acclerating, fastest 4WD you can buy, but the Audi Quattro is no fuel economy champion.

A higher performance version of the Audi 5000 engine, the Quattro's powerplant puts out 160 horsepower, but requires premium gas.

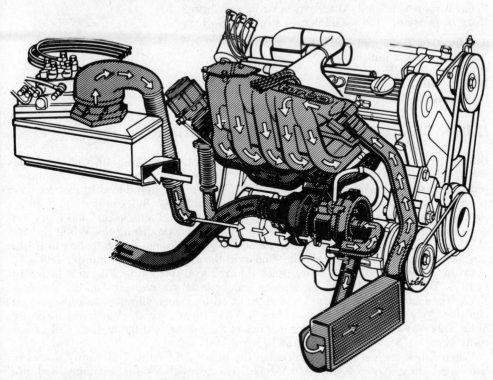

Quattro's intercooler is mounted low. Air is sucked through the fuel injection air flow meter, compressed by the turbocharger, cooled by the intercooler, and pushed through the intake manifold to the cylinders.

Model: Audi Quattro

Importer: Porsche & Audi Division
Volkswagen of America
Englewood Cliffs, NJ 07632

Model year (s):   1983—(continuing)
Specifications for: 1983 Audi Quattro
Vehicle type: front engine, 4 wheel drive
Body styles: 2-door sedan

Wheelbase: 99.5 in
Length: 178.2 in
Width: 67.9 in
Height: 52.0 in
Weight: 3040 lbs
SAE volume, interior/trunk: 84/8
Fuel tank capacity: 23.8 gal

Engine type: 5 in-line turbo gasoline
Displacement: 2144cc/130.8 cu in
Turbo: KKK with separate wastegate and intercooler
Maximum boost: 12.3 psi
Compression ratio: 7.0:1
Fuel system: Bosch K-Jetronic
Horsepower: 160 @ 5500 rpm

Transmission (s):   5-speed manual

Front suspension: ind, MacPherson strut, coil springs
Rear suspension: ind, MacPherson strut, coil springs

Acceleration, 0–60 mph: 7.8 sec
Top speed: 122 mph
Roadholding: .74g
Braking, 70–0 mph: 200 ft
EPA Fuel economy, city/hwy: 17/28

BUICK CENTURY/LESABRE TURBO COUPE

Buick's turbo V-6 was available in the LeSabre and Century from 1978 to 1980. This engine is a distant cousin of the first production turbocharged V-8, the Oldsmobile Jetfire F85 with a turbocharged aluminum 215 V-8. Buick introduced its cast iron V-6 version of the engine in 1962, with 198 cubic inches. In '64, Buick upped the displacement to 225 cubic inches. Then came the years of the V-8 muscle car, and fuel economy was unimportant to most buyers so Buick stopped producing the V-6 at the end of '67 and sold the rights to Jeep. Oldsmobile sold the aluminum V-8 tooling to British Leyland, who subsequently used the engine in the Rover V-8 and Triumph TR8.

In the interest of fuel economy, Buick reintroduced the V-6 in 231 cubic inch form in 1975. The engine still used the uneven firing crankshaft, derived from the V-8, until 1977, when the even firing crank was introduced to reduce vibration and increase engine life. When Buick turbocharged the V-6 for '78, it became clear that this engine of the past was to be one of Buick's engines of the future, and the high tech electronic controls of the '84 V-6 turbo certainly bear this out.

When Buick engineers began to design the turbo V-6 for the '78 Century, LeSabre, and Regal, they wanted to combine V-8 performance with V-6 fuel economy, and minimize turbo lag. Keep in mind that this turbo development took place in 1977, when the only production turbocharged cars available in America were the Porsche and Saab,

Buick LeSabre Turbo has more interior room than any turbo sedan ever made.

Buick Century Turbo has angular styling.

and virtually no one even knew what a wastegate was. To reduce turbo lag, Buick first tried several turbo layouts and turbocharger placements. Each successive design reduced intake system volume, with a corresponding reduction in lag time. In the final draw-through configuration the compressor housing bolted directly to the intake manifold, and the carburetor is mounted on an adapter which bolts directly to the compressor inlet.

Buick's second step towards minimizing lag was selecting the best turbocharger size. The engineers compared compressor maps, test drove vehicles with various turbochargers, and ran engines with various turbochargers on dynamometers before selecting an AiResearch TB03 with a low .82 A/R ratio. This A/R ratio offered high boost pressures at low engine speed without unnecessary exhaust back pressure.

Buick did not automatically decide to use a wastegate. The desired turbocharger gave good low and mid-rpm power, but uncontrolled, boost pressure would exceed 15 psi at high rpm. Of course, running the 8:1 compression ratio and spark advance Buick wanted for fuel economy with 87 pump octane gasoline, this combination meant detonation. Retarding spark timing severely not only hurt power and fuel economy, but also raised exhaust gas temperatures unacceptably. Increasing boost pressure from seven to nine psi increased horsepower, but when engineers tried 11 psi, power between 2600 and 3800 rpm fell, due to the retarded ignition timing needed to control detonation.

So Buick engineers realized they had to hold boost pressure at nine psi. They didn't know how best to accomplish this. One approach was to change the size of the turbocharger's exhaust nozzle passage while utilizing the same turbine. This is known as the A/R ratio. They didn't want to raise the turbocharger's A/R ratio, because then the engine could not achieve maximum boost at low engine speeds. The engineers tried a fixed exhaust restriction, but environmental changes (altitude, barometric pressure, temperature) created excessive variations in boost pressure, and boost pressure suffered at low engine speeds. A variable inlet restrictor gave the right boost pressure curves, but the fuel/air mixture's temperature increased by as much as 80°F, causing detonation. Finally Buick engineers tried a wastegate, and found it controlled boost pressure without side effects.

Next Buick had to design an ignition timing system that would allow optimum ignition advance under light load conditions, yet retard timing enough during boost to curb detonation. Testing showed that the amount of ignition retard required to curb detonation varied greatly due to atmospheric conditions, fuel octane, car—to—car variances in compression ratio and fuel/air mixture, and combustion chamber temperature. Combustion chamber temperature was found to be particularly significant, so that more advanced timing, resulting in more power, could be tolerated during short bursts of acceleration, but not during sustained high speeds. As a result of these variables, the engineers decided that boost activated retard and double-acting advance/retard distributor systems were unacceptable.

Instead, Buick and Delco envisioned and then developed the revolutionary detonation sensor-equipped electronic spark control system. This system allows normal centrifugal and vacuum advance when no detonation is present, and retards the timing only when detonation occurs. A detonation sensor detects the presence and intensity of all engine vibrations and sends this information to the Turbo Control Center. The early detonation sensors are magnetostrictive devices consisting of a high-nickel-alloy —core permanent magnet and a coil. Vibrations are transmitted to the core, causing a flux density change which generates a voltage in the coil. That is, motion in the core makes a generator of the assembly. Now, simpler, more accurate piezoelectric crystal detonation sensors are used. In either case, the detonation sensor is mounted on the intake manifold in a position over the thermostat which transmits the vibrations from each cylinder with equal intensity.

The detonation sensor reacts to many vibrations, including those caused by the valvetrain, etc., but this normal voltage is rejected as background noise by the Turbo Control Center.

ENGINE/DETONATION SENSOR VIBRATION SPECTRUM

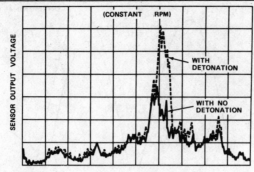

Detonation causes the sensor's voltage to jump dramatically, frequently doubling the voltage created by background noise.

The Turbo Control Center is an electronic module which compares the strength of the detonation induced signal (when present) to the background noise to determine the intensity of the detonation. The Control Center then orders the ignition module in the HEI (high energy ignition) distributor to retard the timing an appropriate amount.

A number of internal engine components were also redesigned for turbo duty. Torque below 2400 rpm was increased by reducing the camshaft's intake lift from .383 to .323" for increased swirl and better breathing at these rpms. The piston domes were made 50% thicker, the second ring land was thickened by .025", and the third ring land by .015". Piston diameter at the ring lands was reduced by .010" to eliminate ring land–to–cylinder wall contact under high heat conditions. The piston–centering boss on the center of the dome was eliminated to reduce the likelihood of preignition from concentrated heat. The turbo piston skirt provides a 30% wider contact area with the cylinder wall to improve heat transfer and to reduce cylinder wall scuffing by spreading the load over a wider area. The second compression and oil scraper rings are chrome plated for reduced wear in the turbo engine. (The top ring was already Moly–filled, and rugged enough to take the extra stress.)

Early dyno testing caused cracks in the fillets of the crankshaft. A rolled-fillet process of manufacturing, which induces static compressive stress to reduce flexing under load, was substituted for the conventional fillet machining to eliminate these cracks. This rolled fillet process increased crankshaft bench fatigue life five-fold.

In 1978 two-barrel and four-barrel versions of the V-6 turbo were available, but for '79 the two-barrel was dropped. High speed vehicle testing in '77 revealed that both the Rochester Dualjet two-barrel and Quadrajet four-barrel carburetors ran too lean from three-fifths to four-fifths throttle. This happened because the draw-through carbs received vacuum indicative of a light throttle load, closing the power valves, while the

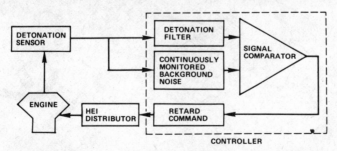

Detonation sensor responds to all engine vibrations, but detonation is more intense, causing the sensor to respond with far greater output voltage.

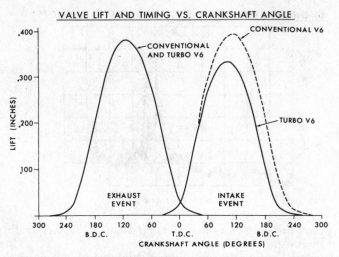

Early Buick V-6 turbo camshaft has less intake valve lift than normally-aspirated cam, improving low rpm torque.

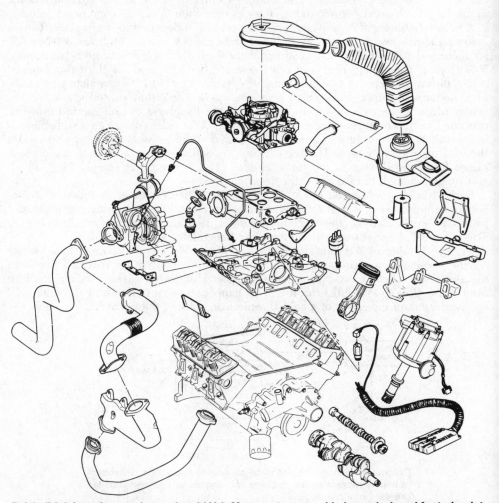

Buick didn't just slap a turbo on the old V-6. Many parts were added or redesigned for turbo duty.

engine was actually under high load due to near—maximum boost pressure. Buick's answer was an external control for the carbs' power enrichment systems. This control senses intake manifold pressure/vacuum. A vacuum signal is routed through this control valve to the carburetor enrichment piston. But at pre-set manifold pressure the signal is bled to atmosphere, opening the power valves and causing enrichment. The fuel/air mixture at full throttle is not affected since power valves are open then anyway.

The torque converter's stall speed was upped from 1400 to 2000 rpm to build boost quicker during standing start acceleration. The converter's diameter was increased from 11.75″ to 12.2″ to reduce wasteful slippage.

Acceleration of the V-6 4-barrel turbo-equipped '78 Buicks was on par with those powered by the 350 4-barrel V-8, with two mpg better EPA gas mileage than the V-8. Fuel economy fell short of the normally-aspirated V-6 by one mpg with the two-barrel turbo engine and two mpg with the four-barrel turbo setup. But while the normally aspirated V-6 put out 105 horsepower, the two-barrel turbo had 150 horsepower and the four-barrel turbo 165. The turbo V-6 weighs 51 pounds more than the normally-aspirated V-6, and 134 pounds less than the big V-8.

Model: Buick Century Turbo Coupe

Manufacturer: General Motors Corp.
                Buick Motor Div.
                Flint, MI 48550

Model year (s):   1979–80
Specifications for: 1980 Buick Century Turbo Coupe
Vehicle type: front engine, rear drive
Body styles: 2-door sedan, 4-door sedan

Wheelbase: 108.1 in
Length: 196.0 in
Width: 71.6 in
Height: 54.1 in
Weight: 3500 lbs
SAE volume, interior/trunk: 97/16
Fuel tank capacity: 22.0 gal

Engine type: V-6 turbo gasoline
Displacement: 3791cc
Turbo: AiResearch TB03 with integral wastegate
Maximum boost: 9 psi
Compression ratio: 8.0:1
Fuel system: 4-barrel Rochester carburetor
Horsepower: 165 @ 4000 rpm

Transmission (s):   3-speed auto

Front suspension: ind., control arm, coil springs
Rear suspension: rigid axle, coil springs

Acceleration, 0–60 mph: 10.0 sec
Top speed: 105 mph
Braking, 70–0 mph: 220 ft
EPA Fuel economy, city: 18

Model: Buick LeSabre Turbo Coupe

Manufacturer: General Motors Corp.
                Buick Motor Div.
                Flint, MI 48550

Model year (s):   1979–80
Specifications for: 1980 Buick LaSabre Turbo Coupe
Vehicle type: front engine, rear drive
Body styles: ·2-door sedan, 4-door sedan

Wheelbase: 115.9 in
Length: 218.4 in
Width: 75.9 in
Height: 55.0 in
Weight: 3800 lbs
SAE volume, interior/trunk: 107/21
Fuel tank capacity: 22.0 gal

Engine type: V-6 turbo gasoline
Displacement: 3791cc
Turbo: AiResearch TBO3 with integral wastegate
Maximum boost: 9 psi
Compression ratio: 8.0:1
Fuel system: 4-barrel Rochester carburetor
Horsepower: 165 @ 4000 rpm

Transmission (s):   3-speed auto

Front suspension: ind., control arm, coil springs
Rear suspension: rigid axle, coil springs

Acceleration, 0–60 mph: 10.8 sec
Top speed: 101 mph
Braking, 70–0 mph: 230 ft
EPA Fuel economy, city: 16

BUICK REGAL TURBO

For 1984, Buick's V-6 turbo produces 200 horsepower with 13 pounds of boost—in first and second gears only. In third gear, maximum boost drops down to 12 psi and horsepower falls to 195. Finally, in fourth gear, only 8 pounds of boost is allowed, producing 180 horsepower. The purpose of this staggered boost is to allow maximum boost in the lower gears, where there isn't much load and maximum boost is not sustained, while reducing the chance of detonation in the high load, sustained speed-and-boost higher gears—particularly in the automatic transmission's overdrive fourth gear. One of the reasons this system works is that only sustained high load produces maximum combustion chamber temperatures and the related tendency to detonate.

An electronically-controlled wastegate system makes this staggered boost possible. The electronic control module (ECM) signals the wastegate solenoid to bleed a tiny amount of pressurized air from the intake manifold to the wastegate's actuator. The actuator is a diaphragm in a can, and the air pressure in the can pushes the diaphragm, which opens the wastegate via a rod. When the solenoid shuts off air pressure to the actuator, a spring closes the wastegate.

The '84 Buick turbo is one of only four mass-market production car engines which modulate both boost pressure and ignition retard when detonation is present. The other three are the Mustang SVO, Pontiac 2000 Turbo, and Audi 5000S Turbo.

In the '84 Buick Turbo, the ECM (electronic control module) controls ignition timing, boost pressure, fuel/air mixture, idle speed, exhaust gas recirculation, and torque converter lock up. The ignition timing section of the ECM is the EST (electronic spark timing), and it receives signals from the coolant temperature sensor, throttle position sensor, crank position sensor, cam position sensor, mass flow sensor, detonation sensor, and the ECM's internal stopwatch, which gives the amount of time since engine start-up. The detonation sensor's signal runs through a filter called the ESC (electronic spark control), mounted on the right inner fender, before being sent to the EST.

The EST can retard ignition timing by as much as 30° to eliminate detonation, and it does so with incredible speed. The EST receives 80 updates from the detonation sensor per second, and can retard the ignition timing by the full 30° in as little as one-twelfth of a second.

The wastegate control section of the ECM receives data concerning load (from the mass air flow sensor), ignition timing, engine speed, transmission gear, and detonation. While the EST receives an indication of detonation strength from the ESC, the wastegate control section only receives a "yes or no" detonation signal. When sustained detonation is present, the EST first begins to retard the ignition timing. Only when the EST has retarded the timing by 15° does the wastegate control section order the solenoid to begin reducing boost pressure. The wastegate control system can reduce boost pressure from 13 psi to 8 psi in as little as one and a half seconds. Once detonation is eliminated, a different control scheme is utilized. The wastegate control section will not begin to increase boost pressure again until the ignition timing is first advanced to within 15° of its normal setting.

The electronic ported fuel injection system used on the '84 Buick Turbo is timed to inject the fuel sequentially through Bosch injectors into each cylinder's intake port just prior to the opening of that cylinder's intake valve. Each cylinder receives one squirt per four-stroke cycle. The ECM controls the fuel injectors, using information from the coolant temperature sensor, time since engine start-up stopwatch, throttle–position sensor, oxygen sensor, cam and crank position sensors, and the mass air flow sensor.

The '84 Regal Turbo's mass air flow sensor is located in front of the throttle in the air intake. It is a heated film which actually measures the mass of air, rather than the volume. As incoming air passes over this heated film, the cooling effect is measured and sent as input to the ECM. The sensor consists of a screen to break up the air flow, a resistor, the heated film, and an electronic module. The module maintains the temperature of the heated film at 167°F over the temperature of the incoming air. The heated film is a nickel grid coated with Kapton, a high temperature material. If it takes more electrical current to maintain the film at 167°F, then the incoming mass of air has increased. This information is then sent to the ECM.

The throttle body incorporates an idle air control with a bypass channel through which air can flow. A needle opens and closes an orifice in this bypass channel to regulate idle air flow, and the needle is controlled by a stepper motor on orders from the ECM. The '84 Regal Turbo also has a new camshaft with slightly more valve lift and duration.

A Grand National special edition of the '84 Regal T-Type features the turbo V-6, 215/60 radial tires, a sport steering wheel, 94 amp alternator, alloy wheels, and a Lear Siegler interior with front seats embroidered with the NASCAR Grand National logo.

'84 Buick Grand National Regal has turbo V-6 power, front and rear spoilers, a high output alternator, wide radial tires, and alloy wheels.

The Buick Regal Turbo boasts the longest production run of a turbo car made in America. Production started in '78, and the Regal keeps getitng better—and faster.

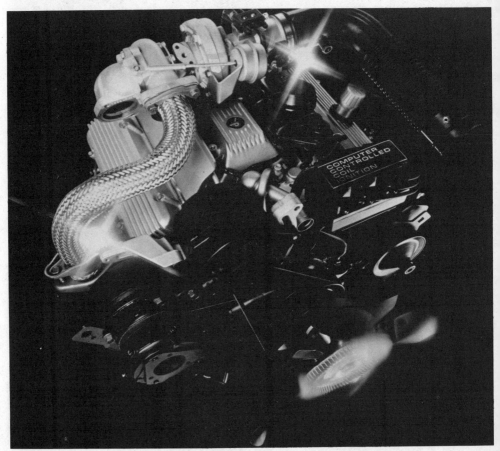

'84 Buick V-6 turbo gives the Regal 200 horsepower. Electronic controls, port fuel injection, and distributorless ignition system are high tech.

The exterior is black from nose to tail: black paint, bumpers, front airdam and rear spoiler.

The Buick Regal Turbo boasts the longest production run of a turbocharged car made in America, having been produced every year since 1978. The engine was originally available in two-barrel (150 horsepower) and four-barrel (165 horsepower) form. In '79 the two-barrel version was dropped. In '81, available horsepower went up to 170, thanks to an optional package with semi-dual exhausts. The package also included a higher stall speed torque converter, and 3.08 (versus 2.73) rear axle ratio.

For '82 horsepower was upped to 175 and throttle response was improved by a turbocharger change to a reduced A/R dimension and an increase in turbine outlet pipe diameter to 2.25 inches, reducing back pressure. The new turbocharger also had an oil squirt hole to bathe the turbine bearing shaft for cooling purposes. Oil capacity was increased from four to five quarts, and a higher output oil pump extended the oil change interval from 3500 to 7500 miles. The '81 performance option became standard in '82. Also in '82 the intake manifold's carb plenum was heated more quickly by exhaust rather than the old style engine coolant system. This heating change permitted shutting off heat to the plenum during full throttle operation for a cooler charge. Also, the early fuel evaporation unit, an electrically heated grid between the carburetor and exhaust manifold which was introduced in 1980, heated faster for '82 to help vaporize fuel during cold starts.

'83 Regal Turbos got stainless steel exhaust manifolds and a larger diameter exhaust crossover pipe, bringing horsepower up to 180. A finned turbocharger housing and heavy-duty radiator kept the engine cool, and the four-speed automatic transmission was introduced with its 3.42 rear axle ratio. With the even greater improvements in the '84, it seems like the Buick Regal's turbo evolution just won't quit.

Model: Buick Regal Turbo

Manufacturer: General Motors Corp.
                  Buick Motor Div.
                  Flint, MI 48550

Model year (s):   1978—(continuing)
Specifications for: 1984 Buick Regal Turbo
Vehicle type: front engine, rear drive
Body styles: 2-door sedan, 4-door sedan

Wheelbase: 108.1 in
Length: 200.6 in
Width: 71.6 in
Height: 55.3 in
Weight: 3580 lbs
SAE volume, interior/trunk: 98/17, 102/16
Fuel tank capacity: 18.1 gal

Engine type: V-6 turbo gasoline
Displacement: 3791cc/231.3 cu in
Turbo: AiResearch TBO3 with electronically-controlled integral wastegate
Maximum boost: 13 psi
Compression ratio: 8.0:1
Fuel system: electronic port fuel injection
Horsepower: 200 @ 4000 rpm

Transmission (s):   4-speed auto

Front suspension: ind., control arm, coil springs
Rear suspension: rigid axle, coil springs

Acceleration, 0–60 mph: 8.0 sec
Top speed: 116 mph
Roadholding: .76g
Braking, 70–0 mph: 202 ft
EPA Fuel economy, city/hwy: 17/28

BUICK RIVIERA TURBO

Since 1979, the Buick Riviera has been the only luxury car available in America with a turbocharged gasoline engine. While its fuel economy is no better than other gasoline-powered luxury cars, like the Lincoln Continental, Oldsmobile Toronado, and Cadillac Eldorado, the Turbo Riviera has always had a horsepower and weight advantage over the competition, yielding better acceleration.

The Riviera was the first American car with front wheel drive and a turbo. The turbocharger on the Riviera has always been mounted at the rear of engine, instead of on top of the engine, as on the Regal. This rear mounting balances the left and right bank exhaust streams, squeezing more work out of the waste exhaust gases than on the Regal. The rear mounting position also reduces intake heating by the engine. As a result, the early Rivieras had a higher horsepower rating than their Regal counterparts. For '84, the Riviera is rated at the same 200 horsepower as the Regal even though the Riviera does not benefit from the Regal's higher lift camshaft. As explained in the section on the Regal, this 200 horsepower rating only applies to first and second gears. The '84 Riviera's turbo V-6 boasts an electronically-controlled wastegate, electronic port fuel injection, and a distributorless ignition system.

The Riviera T-Type comes with the turbo V-6 and Grand Touring suspension as standard equipment. All Rivieras since '79 have featured four wheel independent suspension. The superb Delco-Bose stereo system is available on all Rivieras except the convertible.

The Buick Riviera was the first American car with front wheel drive and a turbo. Today it is still the only turbo gasoline-powered luxury car.

Riviera's turbo is mounted behind the engine, rather than over it, as on the Regal. As a result, this '79 has more power than its Regal counterpart.

'84 Buick Riviera convertible sports a 200 horsepower turbo V-6 for open air motoring with extra punch.

Model: Buick Riviera Turbo

Manufacturer: General Motors Corp.
Buick Motors Div.
Flint, MI 48550

Model year (s):   1979—(continuing)
Specifications for: 1984 Buick Rivera Turbo
Vehicle type: front engine, front drive
Body styles: 2-door sedan, 2-door convertible

Wheelbase: 114.0 in
Length: 206.6 in
Width: 72.8 in
Height: 54.3 in
Weight: 3700 lbs
SAE volume, interior/trunk: 110/16, 90/13
Fuel tank capacity: 21.1 gal

Engine type: V-6 turbo gasoline
Displacement: 3791cc/231.3 cu in
Turbo: AiResearch TB03 with electronically-controlled integral wastegate
Maximum boost: 13 psi
Compression ratio: 8.0:1
Fuel system: electronic port fuel injection
Horsepower: 200 @ 4000 rpm

Transmission (s):   4-speed auto

Front suspension: ind., control arm, torsion bars
Rear suspension: ind., coil springs

Acceleration, 0–60 mph: 8.4 sec
Top speed: 114 mph
Braking, 70–0 mph: 205 ft
EPA Fuel economy, city/hwy: 16/27

CHEVROLET MONTE CARLO TURBO

Chevrolet said: If Buick can have a turbo in the Regal, why can't we have one in the Monte Carlo? And so the Monte Carlo was blessed with an optional 170 horsepower Buick V-6 turbo, upping horsepower by 47% over the base V-6. Yet EPA city fuel economy remained at 19 mpg with the turbo, the same as with the normally aspirated V-6.

The Monte Carlo Turbo was available in '80 and '81, and is easily recognized by the unique hood bulge just to the left of the hood's centerline. For '81, the Monte Carlo received a sheetmetal overhaul with a lower hood and raised rear deck which reduced aerodynamic drag by 10 percent. New options for '81 included alloy wheels and fog lights. In either year the optional F-41 sport suspension boasted 70-series steel belted radials, thicker front and rear anti-sway bars, and stiffer springs and shocks. Limited slip differentials and a T-roof were also optional. The country club's Carlo could get to 60 mph in 10.5 seconds, and go around curves, too! A three-speed automatic is the Turbo's only transmission, and the Computer Command Control system locks up the torque converter while cruising to improve fuel economy.

An AiResearch TB03 turbocharger boosts the intake manifold pressure up to a maximum of 9 psi. The detonation sensor-equipped computer control system retards the ignition timing when detonation is present. More information on this system and the V-6 turbo's development can be found in the Buick section of this chapter.

Chevy's only turbo car, the Monte Carlo has 170 Buick V-6 horsepower. Note the unique hood bulge.

Model: Chevrolet Monte Carlo Turbo

Manufacturer: General Motors Corp.
Chevrolet Motor Div.
Warren, MI 48090

Model year (s):   1980–81
Specifications for: 1981 Chevrolet Monte Carlo Turbo
Vehicle type: front engine, rear drive
Body styles: 2-door sedan

Wheelbase: 108.1 in
Length: 200.4 in
Width: 71.8 in
Height: 53.9 in
Weight: 3200 lbs
SAE volume, interior/trunk: 97/16
Fuel tank capacity: 18.1 gal

Engine type: V-6 Turbo gasoline
Displacement: 3791cc/231.3 cu in
Turbo: AiResearch TB03 with integral wastegate
Maximum boost: 9 psi
Compression ratio: 8.0:1
Fuel system: 4-barrel Rochester carburetor
Horsepower: 170 @ 4000 rpm

Transmission (s):   3-speed auto

Front suspension: ind., control arm, coil springs
Rear suspension: rigid axle, coil springs

Acceleration, 0-60 mph: 10.5 sec
Top speed: 103 mph
Roadholding: .71g
EPA Fuel economy, city/hwy: 19/27

DATSUN 280ZX TURBO

The 280ZX Turbo is the quickest turbocharged car on the market, as well as the quickest car out of Japan. Its 6.8 second 0-60 time can only be beaten by a quintet of high performance domestic V-8s, a couple of Porsches, and a lone Ferrari.

The installation of an AiResearch TB03 turbocharger has brought about this improvement to the ZX's already quick performance. Other changes include a 2.5 inch diameter exhaust system (up from 1.9 inches), a 50 percent larger capacity muffler, different cam timing, a bigger air flow meter and fuel injectors, a higher voltage coil, a bigger radiator, and an oil cooler. Low 7.4:1 compression pistons have improved wrist-pin lubrication, and the oil pump and cylinder head bolts are enlarged. The turbo also gets a NACA duct hood to minimize intake air turbulence.

The ECCS microprocessor system receives information from the crank angle/rpm sensor, water temperature sensor, barometric pressure sensor, air temperature sensor, oxygen sensor, air flow meter, and detonation sensor. Automatic transmission versions also have a park/neutral switch and a car speed sensor. With this information the digital microprocessor controls the electronic ported fuel injection, ignition timing, exhaust gas recirculation, idle speed, and fuel pump operation. The crank angle/rpm sensor works off the front end of the crankshaft. A disc with 93 teeth spins on the end of the crankshaft, and three magnetic sensors use these teeth to generate rpm and crank angle signals for the microprocessor.

Although the 280ZX Turbo sticks well around a smooth curve, it shares with every 280 a nasty trait when driven fast through the twisties. When entering a curve fast, the front end starts to slide. It soon regains its grip, but as soon as it does, the rear end whips out. This transient roll-oversteer has plagued the Z-car since the old days when it used MacPherson struts all around. Today's softly sprung semi-training arm rear suspension, plus the Turbo's extra power, makes its behavior even worse.

For 1983, when most GTs were getting stiffer suspension, the ZX Turbo got even softer. Damping is reduced, and the front springs are softer. An optional leather interior is available for 1983, in an option package which includes a digital dash, air conditioning, bronze tinted glass, electrically defogging outside mirrors, and an uprated

Datsun 280ZX Turbo is the quickest accelerating turbo car on the market.

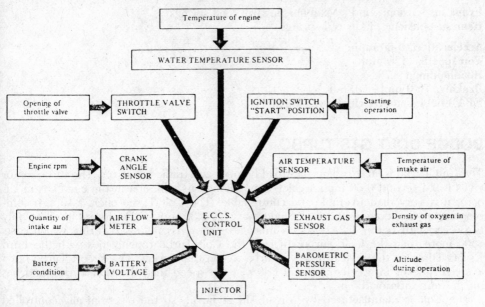

Datsun's ECCS microprocessor control unit is barraged by information.

stereo system. Other options included headlight washers, a T-top, alloy wheels, and a sensor system which monitors lights and fluid levels. The 280 ZX Turbo was available as a two seater or a 2+2, which weighs 110 pounds more.

Model: Datsun 280 ZX Turbo

Importer: Nissan Motor Corp.
       Box 191
       Gardena, CA 90247

Model year (s):   1981–83
Specifications for: 1983 Datsun 280ZX Turbo
Vehicle type: front engine, rear drive
Body styles: 3-door coupe, 3-door sedan

Wheelbase: 91.3 in
Length: 174.0 in
Width: 66.5 in
Height: 51.0 in
Weight: 2990 lbs
SAE volume, interior/trunk: 72/14
Fuel tank capacity: 21.1 gal

Engine type: 6 in-line turbo gasoline
Displacement: 2753cc/168 cu in
Turbo: AiResearch TBO3 with integral wastegate
Maximum boost: 6.8 psi
Compression ratio: 7.4:1
Fuel system: electronic port fuel injection
Horsepower: 180 @ 5600 rpm

Transmission (s):   5-speed manual, 3-speed auto

Front suspension: ind., MacPherson strut, coil springs
Rear suspension: ind., coil springs

Acceleration, 0-60 mph: 6.8 sec
Top speed:. 126 mph
Roadholding: .78g
Braking, 70-0 mph: 196 ft
EPA Fuel economy, city/hwy: 19/28

DODGE COLT GTS TURBO

The Colt Turbo is an excellent backroad bomber and traffic terrorizer. The turbo motor gives it 9.4 second 0-60 mph acceleration. Though its adhesion in the curves is mediocre, it is very maneuverable and controllable. The Colt Turbo offers a superb combination of acceleration and fuel economy, with a 30 mpg EPA city rating. The only turbo car with better EPA city gas mileage is the Nissan Pulsar Turbo, which costs $800 more than the Colt Turbo. In fact, the Colt Turbo roughly ties with the Ford Escort Turbo as the least expensive Turbo car on the market. The Escort Turbo however, is quicker in acceleration and faster in top speed than the Colt Turbo, and has much more interior room.

The Colt is manufactured by Mitsubishi in Japan, so the electronically controlled fuel injection system is similar to those on the Starion, Cordia, and Tredia turbos. Mitsubishi's unique ultrasonic air flow sensor measures intake air flow. Both the air flow sensor and air temperature sensor sit in the air cleaner. The coolant temperature sensor is located in the intake manifold. The throttle position sensor sits at the end of the throttle shaft.

An electric idle speed control motor holds a steady idle by extending or retracting its throttle stop plunger. Unlike the common solenoid-operated idle stop, this unit actually has a rotating motor which spins a worm gear to move the plunger. A sensor signals the idle plunger position, and a separate sensor indicates when the throttle touches the idle stop. A starter motor sensor indicates when the engine is being cranked over by the starter. A distributor-mounted sensor signals engine speed, while a boost pressure sensor is located on the firewall beside the wiper motor, receiving pressure from a hose connected to a nipple on the intake manifold. The detonation sensor sits on the

Dodge Colt Turbo is the epitome of pocket rockets. It's made in Japan by Mitsubishi.

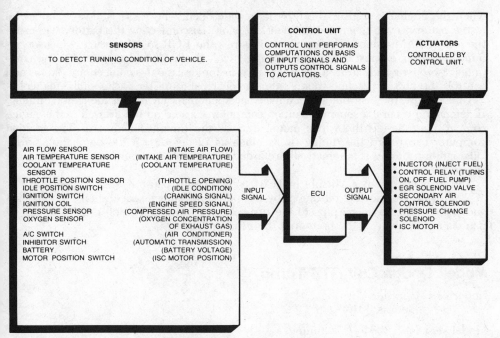

The Colt's engine sensors on the left provide the ECU microprocessor with the information it needs to control the functions on the right.

side of the engine block. An oxygen sensor senses the oxygen in the exhaust after it exits the turbocharger.

On the basis of this information, the computer controls the throttle body fuel injection fuel/air mixture, ignition timing, emission controls, and idle speed. This computer is located on the front passenger's seat cowl panel, and it controls the fuel injection two ways. It controls the quantity of fuel squirted per injection and the number of injections per second. There are two injectors in the throttle body and they alternate in delivering the fuel. The computer also controls the position of the idle speed control motor which sets idle speed.

The computer actually has a RAM (Random Access Memory). When the oxygen sensor feeds the computer information on the fuel/air mixture, the RAM memorizes the latest info to assist the computer in selecting the optimum number of injections per second

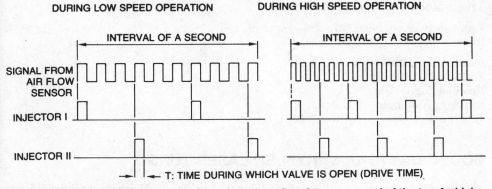

The ECU controls the volume per squirt and number of squirts per second of the two fuel injectors which reside in one throttle body.

when the engine is cold or the throttle is wide open. The RAM is always connected to the battery, even when the ignition key is off. Disconnecting the battery will cause the loss of RAM values. The computer controls the EGR solenoid valve which determines whether exhaust gases will be mixed with the fresh fuel/air mixure. The computer also opens and closes a solenoid valve in the emissions air injection system, and controls the flow of gasoline vapors from the charcoal cannister to the intake manifold.

Under boost, the distributor vacuum advancer automatically retards the ignition timing. If detonation occurs, a separate ignition computer, located on the firewall of the engine compartment beside the wiper motor, retards the timing even more. The ignition computer can retard the timing by as much as 12 degrees. If the detonation sensor fails, ignition timing is automatically retarded by 8 degrees under boost to protect the engine.

The turbocharger is a Mitsubishi TC04, which builds boost at engine speeds as low as 2000 rpm. Lubricating oil drawn by the oil pump branches at the oil filter bracket and passes through an oil pipe to the turbocharger bearing. The oil returns to the oil pan via a drain hole at the bottom of the bearing housing.

Model: Dodge Colt GTS Turbo

Importer: Chrysler Corp.
Detroit, MI 48288

Model year (s): 1984—(continuing)
Specifications for: 1984 Dodge Colt GTS Turbo
Body styles: 3-door sedan

Wheelbase: 90.6 in
Length: 156.9 in
Width: 62.4 in
Height: 50.0 in
Weight: 1950 lbs
SAE volume, interior/trunk: 75/11
Fuel tank capacity: 13.2 gal

Engine type: 4 in-line turbo gasoline
Displacement: 1598cc/97.5 cu in
Turbo: Mitsubishi TC04 with integral wastegate
Maximum boost: 8.7 psi
Fuel system: electronic throttle body fuel injection
Horsepower: 102 @ 5500 rpm

Transmission (s): 2x4-speed manual

Front suspension: ind, MacPherson strut, coil springs
Rear suspension: ind, coil springs

Acceleration, 0-60 mph: 9.4 sec
Top speed: 109 mph
Roadholding: .70g
Braking, 70-0 mph: 208 ft
EPA Fuel economy, city/hwy: 30/37

DODGE DAYTONA/CHRYSLER LASER TURBO

America's first front wheel drive turbocharged sports car is a very good all-around performer. While it doesn't rate "best" in any cateogry, its very good combination of ac-

Dodge's Daytona Turbo is a super looker and an excellent all-around performer.

The Chrysler Laser Turbo offers performance with a touch of class—and a neat digital dashboard.

celeration, fuel economy, top speed, handling, brakes, interior room, good looks, and reasonable price make it an attractive performance car.

Much of that all-around performance is due to the 2.2 liter turbo motor. Its 142 horsepower is 30% greater than the already strong–running Shelby 2.2 motor. The engine boasts port fuel injection, with throttle body and injectors by Bosch, plumbing and electronic controls by Chrysler. Chrysler's computer makes 100 calculations per second in order to properly adjust the fuel/air mixture and ignition timing. This computer is served by eight sensors: throttle position (or opening angle), boost pressure, engine rpm, engine coolant temperature, fuel/air mixture temperature, distributor position, exhaust oxygen, and detonation. Idle speed is held constant by an automatic idle speed control. The computer even has a learning circuit that remembers what it had to do the last time to get the proper exhaust oxygen content (indicating the proper fuel/air mixture) and corrects any earlier mistakes in fuel scheduling. Premium gas is recommended, but the detonation sensor backs off the ignition timing when regular unleaded is used.

The AiResearch T03 turbocharger builds 7.2 psi boost at a low 2050 rpm, with a maximum boost of 7.5 psi at higher engine speeds. Chrysler's version of the AiResearch turbo boasts a unique watercooled bearing housing which reduces peak operating temperatures. The spinning shaft inside a turbocharger can suffer from hot shutdown failure when the engine is turned off after hard use. Oil pressure drops but the shaft keeps spinning for a few seconds, which can cause bearing damage. The watercooled bearing reduces the probability of bearing damage by removing heat produced by both friction and conduction from the hot turbocharger turbine. After all, Dodge and Chrysler have to back up their 5 year/50,000 mile powertrain warranty.

Also for reliability, the 2.2 turbo motor has high temperature valves, better–sealing piston rings, and a larger capacity oil pump. The camshaft has less overlap and dura-

Chrysler's turbo 2.2 puts out 142 horsepower. A watercooled turbocharger bearing housing extends bearing life for Chrysler's 5 year/50,000 mile powertrain warranty.

tion to improve torque characteristics, and dished pistons lower the compression ratio to 8.2:1.

The Daytona/Laser offers the same interior room and much more trunk space than the Mustang, Camaro and Firebird in a package that is 4 inches shorter than a Mustang, 13 inches shorter than a Camaro, and 15 inches shorter than a Firebird. While the Camaro Z-28, Firebird Trans-Am and Mustang 302GT offer even higher performance than the Daytona Z, the Daytona gets 5 mpg better EPA city gas mileage than they do. The Mustang SVO offers better acceleration and top speed with similar gas mileage to the Daytona, but the SVO costs 50% more. Of course, when Dodge really wants to blow off the performance competition, they'll simply drop the turbo motor into the Charger, which is 430 lbs. lighter.

Model: Dodge Daytona Z/Chrysler Laser Turbo

Manufacturer: Chrysler Corp.
            Detroit, MI 48288

Model year (s):   1984—(continuing)
Specifications for: 1984 Dodge Daytona Z/Chrysler Laser Turbo
Vehicle type: front engine, front drive
Body styles: 3-door sedan

Wheelbase: 97.1 in
Length: 175 in
Width: 69.3 in
Height: 52.0 in
Weight: 2830 lbs.
SAE volume, interior/trunk: 84/17
Fuel tank capacity: 14.0 gal

Engine type: 4 in-line turbo gasoline
Displacement: 2213cc/135 cu in
Turbo: AiResearch T03 with integral wastegate
Maximum boost: 7.5 psi
Compression ratio: 8.2:1
Fuel system: electronic port fuel injection
Horsepower: 142 @ 5600 rpm

Transmission (s):   5-speed manual, 3-speed auto

Front suspension: ind., MacPherson strut, coil springs
Rear suspension: rigid axle, coil springs

Acceleration, 0-60 mph: 8.2 sec
Top speed: 122 mph
Roadholding: .78g
Braking, 70-0 mph: 204 ft
EPA Fuel economy, city/hwy: 22/38

DODGE 600/CHRYSLER LEBARON TURBO

In normally-aspirated form, the Dodge 600 is one of the quickest cars of its interior size. It also gets the best EPA city mileage of any car of its interior size. The turbo drops EPA city economy from 23 to 20 mpg, and EPA highway fuel economy from 38 to 34 mpg. But with a 142 horsepower turbo motor replacing the 98 horsepower normally-aspirated engine, 0-60 mph acceleration drops from 11.4 seconds to a quick 8.6 seconds. This kind of acceleration hasn't been available in a sedan of the Dodge 600's interior room since the '60s—and big motor Pontiac Catalinas didn't get half the Dodge 600's fuel economy.

Premium gas is recommended with the turbo, though a detonation sensor reduces the ignition advance if regular unleaded is used. The engine sports port fuel injection and a watercooled turbocharger bearing housing. And all that power is protected for 5 years or 50,000 miles (whichever comes first), thanks to Chrysler's powertrain warranty.

For optimum handling, select the 600ES. The ES gets Goodyear Eagle GT tires, stiffer front and rear sway bars, and better shock absorbers with more dampening. The

Dodge's 600ES offers lots of interior room, excellent fuel economy, and quick acceleration. The ES handling package includes Eagle GT tires, quicker ratio steering, and heavy duty anti-sway bars and shocks.

Chrysler LeBaron Turbo is respectable enough for bankers—but it's quick! Base springs are soft, so the SDB heavy duty suspension option is to be carefully considered.

Dodge 600 Turbo convertible allows open air turbo pleasure.

power steering ratio is quickened from 17:1 to 14:1. Those who like to turbo around under the sun and stars can select the Chrysler LeBaron turbo convertible. The convertible is 120 pounds heavier and an inch and a half taller than the hardtop, reducing performance and fuel economy slightly.

Model: Dodge 600/Chrysler LeBaron Turbo

Manufacturer: Chrysler Corp.
Detroit, MI 48288

Model year (s):   1984—(continuing)
Specifications for: 1984 Dodge 600/Chrysler LeBaron Turbo
Vehicle type: front engine, front drive
Body styles: 3-door sedan, 4-door sedan, 2-door convertible

Wheelbase: 100.3 in
Length: 178.8–187.2 in
Width: 68.4 in
Height: 52.5–54.1 in
Weight: 2600–2700 lbs
SAE volume, interior/trunk: 97/17
Fuel tank capacity: 14.0 gal

Engine type: 4 in-line turbo gasoline
Displacement: 2213cc/135 cu in
Turbo: AiResearch T03 with integral wastegate
Maximum boost: 7.5 psi
Compression ratio: 8.2:1
Fuel system: electronic port fuel injection
Horsepower: 142 @ 5600 rpm

Transmission (s):   5-speed manual, 5-speed auto

Front suspension: ind., MacPherson strut, coil springs
Rear suspension: rigid axle, coil springs

Acceleration, 0-60 mph: 8.6 sec
Top speed: 110 mph
Roadholding: .76g
Braking, 70-0 mph: 213 ft
EPA Fuel economy, city/hwy: 20/34

FIAT SPIDER/AZZURA TURBO

Introduced in 1981, Fiat's Spider Turbo was the first production turbocharged convertible in America . . . if you consider it a production car. The Spider Turbos were shipped to the U.S. as normally-aspirated Spiders. Legend Industries installed the turbo systems at the dock under contract for the American Fiat distributor.

The Fiat Spider Turbo set another first for a production car in America (again assuming you consider the Fiat Turbo a production car): the use of an IHI turbocharger. The compact IHI RHB6 weighs only 13 pounds with its integral wastegate. Legend mounts the turbo tightly against the right side of the engine block on a special annealed cast iron exhaust manifold. Exhaust gases flow out the turbine via a cast outlet elbow that also contains an oxygen sensor. Fresh air first goes through the stock Bosch L-Jetronic air meter, which retains its original location. After being compressed by the turbo-

The American Fiat distributor turbocharged normally aspirated Spiders at the dock to produce the first production turbocharged convertible in America.

charger, the air is piped through a cast aluminum tube over the valve cover and into the stock fuel injection plenum and throttle body. An electric fan and boost gauge are also included.

This turbo system only increases peak horsepower by 18%, up to 120 horsepower. But the boost begins as low as 1400 rpm, peaking at 6 psi at 3050 rpm. Detonation is avoided by the mild boost pressure, plus a staged boost-activated fuel enrichment and ignition retard system. To protect the engine from wastegate failure or any modification which would create more than 6 psi boost, an excess boost sensor momentarily shuts down the fuel injection until the boost pressure drops down.

Both 5-speed manual and 3-speed automatic prototypes were built. The automatic was actually a half second quicker in 0-60 mph acceleration, because the 5-speed's close ratios didn't take advantage of the turbo's wide power band: the boost goes away and must rebuild each time you shift. But Fiat Motors of America anticipated so little demand for the turbo automatic, they couldn't justify the expense of EPA certification. So the automatic version was never sold.

Cromodora alloy wheels and a small front spoiler were standard on the Turbo. Factory options included power windows and leather interior.

Model: Fiat Spider/Azzura Turbo

Manufacturer: Fiat Motors of North America
777 Terrace Ave.
Hasbrouck Heights, NJ 07604

Model year (s):   1981—(continuing)
Specifications for: 1981 Fiat Spider/Azzura Turbo
Vehicle type: front engine, rear drive
Body styles: 2-door convertible

Wheelbase: 89.7 in
Length: 163.0 in
Width: 63.5 in
Height: 48.2 in
Weight: 2400 lbs
Fuel tank capacity: 11.4 gal

Engine type: 4 in-line turbo gasoline
Displacement: 1995cc/121.7
Turbo: IHI RHB6 with integral wastegate
Maximum boost: 6 psi
Compression ratio: 8.1:1
Fuel system: Bosch L-Jetronic electronic port fuel injection
Horsepower: 120 @ 6000 rpm

Transmission (s):   5-speed manual

Front suspension: ind., control arms, coil springs
Rear suspension: rigid axle, coil springs

Acceleration, 0-60 mph: 9.2 sec
Top speed: 104 mph
Roadholding: .72g
Braking, 70-0 mph: 215 ft.
EPA Fuel economy, city/hwy: 24/33

FORD ESCORT GT/MERCURY LYNX RS TURBO

Into pocket rockets? Want to leave Rabbit GTIs so far behind from the stoplight that they can't even read the "TURBO GT" graphics stretched across your rear spoiler?

Ford Escort Turbo GT boasts better fuel economy and more interior room, than a Rabbit GTi.

Want to leave the Rabbit's normally apsirated 1800 gasping for breath on the thruway, with a twelve mile per hour margin? Knowing all the while, in case your conscience is bothering you, that you're getting better gas mileage (by 4 mpg, EPA city)? And still want to enjoy nine cubic feet more interior room than that Autobahn flier, and three cubic feet more room under the hatch?

Sounds like Ford's Escort Turbo GT is just your ticket. Ford has done a remarkable job with its 1.6 liter turbo engine, extracting 115 horsepower from an engine which only put out 73 horsepower in normally aspirated trim. The turbocharger is a blow-through type, located upstream of all fuel metering components of the multiple-port electronic fuel injection system. Because the turbocharger system handles only air instead of air and fuel together, cylinder-to-cylinder air/fuel mixture variations are minimized and the combustion process is optimized for all types of operating conditions.

The small size of the IHI turbocharger allows it to be placed in an optimum underhood position relative to the engine. Because both the compressor and turbine wheels inside the turbocharger are small and light, throttle response and boost accumulation take place rapidly to virtually eliminate "turbo lag" and enhance acceleration and everyday driveability.

Actual turbocharger output is governed by an integral wastegate which controls the passage of exhaust gas past the turbine. The turbocharger is mounted to a nodular cast iron exhaust manifold which is internally split, or bifurcated, to provide "tuned" exhaust gas pulses to the turbine, based on the firing order of the four engine cylinders. This lets the turbine pull more power from the exhaust. The remainder of the exhaust system is of a low backpressure design to aid engine power output.

A high-flow/low-restriction air cleaner with 50 percent more capacity than that of the normally aspirated 1.6 liter engine is used with the turbocharger system. It includes a special filter element with no vinyl edges to restrict air flow. The vane-type airflow meter and the intake manifold are increased in size over the 1.6 liter engine's components to accommodate the increased airflow provided by turcocharging. And the fuel injectors at each cylinder's intake port are engineered to deliver more fuel when needed for high turbo–boost power output.

The sophisticated EEC-IV electronic engine control system used with the 1.6 liter EEI Turbo engine provides precise control of fuel injection in response to throttle position, air temperature, engine temperature, altitude and engine emissions levels. It also controls spark timing and exhaust-gas recirculation events, and automatically shuts off the air conditioner compressor at wide-open throttle to eliminate that source of power

1.6 liter port fuel injected Escort engine puts out 115 horsepower, 57% more than the normally aspirated version.

drag when full engine output is demanded. Still other control functions include deceleration fuel shutoff, idle speed, evaporative emissions, fuel pressure, and cold starting and cold running fuel enrichment. It also is capable of performing self-test and engine diagnosis functions.

Engine compression ratio is 8.0 to 1, lowered from the 9.5 to 1 ratio of the non-turbocharged 1.6 liter EFI engine. Pistons are of forged aluminum with revised ring groove locations to promote a more effective compression seal. Piston heads are machined to provide a precise, close squish height to generate swirl in the combustion chamber and minimize detonation at all rpm ranges. An on-center piston pin location reduces friction against the cylinder walls.

A high-lift overhead camshaft having the same characteristics as the camshaft of the 1.6 liter HO engine is used to aid engine breathing. Valve springs also have increased tension to ensure they follow cam profile precisely, even under conditions of high engine speeds and high boost pressure (which tends to hold intake valves open). To handle the higher internal heat brought about by turbocharger operation, the engine includes a built-in watercooled oil cooler for crankcase oil. The engine cooling system has twin 60-watt electric, 10-inch, four-bladed fans mounted ahead of the radiator so cooling air is pushed through the fins.

The 1.6 liter EFI turbocharged engine is mated to the five-speed manual transaxle with an overdrive top gear that has a 0.75 to 1 equivalent ratio. The Turbo package also includes TR aluminum wheels and wider 185/65R365 Michelin TRX tires. Major revisions to the TR suspension system include Koni shocks with unique valving, stiffer front springs and a lowering of the front ride height by three-quarters of an inch.

Despite these changes, handling and braking are the Escort's weak points. While not actually inferior to many other cars in its class, the handling and braking ability of the Escort Turbo GT are not up to the level of excellence one would hope for in one of the quickest accelerating–cars you can buy.

Model: Ford Escort Turbo GT/Mercury Lynx Turbo RS

Manufacturer: Ford Motor Co.
Rotunda Dr.
Dearborn, MI 48121

Model year (s): 1984—(continuing)
Specifications for: 1984 Ford Escort Turbo GT/Mercury Lynx Turbo RS
Vehicle type: front engine, front drive
Body styles: 3-door sedan

Wheelbase: 94.2 in
Length: 163.9 in
Width: 65.9 in
Height: 53.3 in
Weight: 2110 lbs
SAE volume, interior/trunk: 85/17
Fuel tank capacity: 13.0 gal

Engine type: 4 in-line turbo gasoline
Displacement: 1598 cc/97.9 cu in
Turbo: IHI with integral wastegate
Maximum boost: 8 psi
Compression ratio: 8.0:1
Fuel system: electronic port fuel injection
Horsepower: 115 @ 5200 rpm

Transmission (s): 5 speed manual

Front suspension: ind. MacPherson strut, coil springs
Rear suspension: ind. MacPherson strut, coil springs

Acceleration, 0–60 mph: 7.9 sec
Top speed: 116 mph
Roadholding: .78g
Braking, 70–0 mph: 210 ft
EPA Fuel economy, city/hwy: 29/42

FORD EXP TURBO COUPE

Here we have Ford's top speed wonder car. The EXP Turbo Coupe is tied neck and neck with Ford's own Mustang SVO as the fastest turbocharged car you can buy; top speed is 128 mph. How can the 115 horsepower EXP motor dream of keeping up with the 174 horsepower SVO? Aerodynamics, my friend. Aerodynamic drag is composed of two factors: drag coefficient (a measure of the shape's sleekness) and frontal area (the widest cross-sectional area of the car that faces the wind). The EXP's drag coefficient is less than the SVO's: .36 versus .41, and the narrow EXP has a much smaller frontal area to push against the wind. The EXP is six inches narrower and over an inch lower than the SVO. While the SVO requires 72 horsepower to overcome aerodynamic drag at 100 mph, the EXP needs only 42 horsepower. 128 mph out of 1.6 liters. And to think that back in '77, Ford's fastest car was a 6.6 liter Thunderbird—at 111 mph. That's progress!

Though the EXP enjoys a big aerodynamic advantage over its family-sized cousin Escort, the EXP is a tad slower in low speed acceleration, due to 140 pounds of extra weight. That's right, the two seat EXP is heavier than the five passenger Escort.

Unfortunately, the EXP Turbo's brakes were not improved to handle 26 mph higher top speed. The EXP's braking system was mediocre before, and high speed driving can cause some exciting moments.

The Turbo Coupe comes with wider 185/65R365 Michelin TRX tires, power steering, Koni shocks, stiffer front springs and a three-quarter inch lower front ride height, which reduces torque steer during low speed acceleration by reducing the half-shaft angles. The EXP's handling is steady and confidence-inspiring, though it's ultimate roadholding is only average.

Aerodynamics and horsepower are the keys to top speed, and the Ford EXP Turbo has both. Top speed is 128 mph, yet the EXP Turbo yields 26 EPA city miles per gallon.

The 1.6 liter EXP Turbo features forged pistons, an oil-to-water oil cooler, and two electric fans.

Model: Ford EXP Turbo Coupe

Manufacturer: Ford Motor Co.
Rotunda Dr.
Dearborn, MI 48121

Model year (s): 1984—(continuing)
Specifications for: 1984 Ford EXP Turbo Coupe
Vehicle type: front engine, front drive
Body styles: 3-door coupe

Wheelbase: 94.2 in
Length: 170.3 in
Width: 63.0 in
Height: 50.5 in
Weight: 2250 lbs
SAE volume, interior/trunk: 75/25
Fuel tank capacity: 13.0 gal

Engine type: 4 in-line turbo gasoline
Displacement: 1598cc/97.5 cu in
Turbo: IHI with integral wastegate
Maximum boost: 8 psi
Compression ratio: 8.0:1
Fuel system: electronic port fuel injection
Horsepower: 115 @ 5200 rpm

Transmission (s): 5-speed manual

Front suspension: ind., MacPherson strut, coil springs
Rear suspension: ind., MacPherson strut, coil springs

Acceleration, 0–60 mph: 8.1 sec
Top speed: 128 mph
Roadholding: .75g
Braking, 70–0 mph: 210 ft
EPA Fuel economy, city/hwy: 26/40

FORD MUSTANG/MERCURY CAPRI TURBO

Back in 1979, Ford introduced their original Mustang/Capri turbo. Based on their 86 horsepower normally-aspirated 2.3 liter four cylinder, this engine put out 130 horsepower. It used a draw-through design with a Holly 6500 two-barrel carburetor, compressed the fuel/air mixture into a conservative six pounds of boost, and used a high 9:1 compression ratio. The AiResearch T03 turbocharger with its integral wastegate was bolted to a special aluminum intake manifold.

The 2.3 liter was considerably beefed up for turbo duty. The cylinder head gasket was changed from perforated core to solid metal core with RTV bead in critical areas. Silicon-chromium steel intake valves and composite nickel-chromium head, silicon chromium steel stem exhaust valves were used. Pistons were forged. A high load oil pump relief spring bumped oil pressure up from 50 psi to 55 psi @ 2000 rpm. A larger oil pan increased total oil capacity by a half quart to five and a half quarts. Main and connecting rod bearings were improved.

Despite these heavy duty components, the engine's reliability record combined with detonation and driveability problems experienced by some owners was not to Ford's liking. When, at the end of 1980 the engine failed to meet the stiffer 1981 emission standards, Ford discontinued it.

It seemed that turbocharging was dead at Ford after this experience, but in late '83 Ford introduced an all new 2.3 liter turbo engine for the Mustang and Capri. And for 1984, Ford has more turbocharged models (fourteen to be exact) available in the U.S. than any other manufacturer. The turbo seed does not die easily—it just awaits re-planting.

The new 2.3 liter turbo uses completely different technology from the old version. The old turbo used a relatively high 9:1 compression ratio, while the new version uses a lower 8:1 ratio. The old version used only six pounds of boost, while the new uses a much higher ten pounds. The draw-through system has been replaced by a blow-through design and carburetion has been replaced by electronic port fuel injection.

1980 Mercury Capri Turbo was good looking, but reliability problems caused Ford to discontinue it at the end of the year.

The '84 Capri Turbo has a bubble rear hatch and a completely redesigned turbo system.

Ford Mustang Turbo GT is even available as a convertible.

The fuel injection system is controlled by the EEC-IV engine control computer using various sensing elements. Among these elements is a vane-type airflow meter located in the path of the air entering the turbocharger. The meter consists of a doorlike vane held closed by light spring pressure. The force of the airflow causes the vane to open against the pressure of the spring proportionately to the amount of air enterig the engine.

A position sensing device on the vane shaft transmits a signal to the electronic control unit that reflects the amount of air entering the engine. The electronic fuel injection system then delivers the proper quantity of fuel required for only the amount of fresh air drawn into the engine. The control of fuel from the injectors by the electronic control unit is practically instantaneous with any change in airflow. Any supplementary-mixture enrichment required as the throttle is opened (as provided by a carburetor's accelerator pump) is supplied through signals from the electronic control unit. Idle stability is also enhanced.

Engine output is controlled by a throttle plate downstream from the airflow sensor in the same manner as a carburetor's throttle plate. Throttle position is controlled by a linkage from the driver's accelerator pedal, as with a conventional carbureted engine. A sensor on the throttle shaft relays throttle position information to the electronic engine control unit.

The new Ford 2.3 liter turbo has port fuel injection, a lower compression ratio, and more boost pressure. Horsepower is up to 145.

During idle, the throttle plate is closed and idle airflow is through an electrically-operated throttle air by-pass valve on the throttle body. This valve also controls airflow for cold-start fast idle, air conditioning-on idle, and closed throttle deceleration.

Another sensor feeding data to the electronic engine control unit is a barometric pressure sensor mounted in the engine compartment. It provides a signal used by the EEC-IV system to modify the fuel/air mixture and spark advance in accord with ambient air pressure to compensate for changes in altitude when the car is driven at varying elevations above sea level.

Still other engine sensors include one that measures the intake air temperature, one that measures engine coolant temperature, and an exhaust gas sensor that detects the amount of excess oxygen remaining in the hot exhaust gases. In addition, a sensor on the distributor housing provides crankshaft position information and engine speed input.

The EEC-IV system uses input from these sensors to open and close the electromagnetic pintle in each fuel injector. The amount of fuel discharged through the orifice of each injector is accurately controlled by regulating the length of time the injector valve remains open.

The new turbo Mustang/Capri is only available with a five speed. For 1984, the heavy duty suspension includes nitrogen shocks and struts. A limited slip rear axle is standard. A front air dam with fog lights comes on the Capri Turbo RS, and will be available in late '84 on the Mustang Turbo GT. The Capri has a bubble-glass rear hatch and wheel flairs which make it 1.7 inches wider.

Model: Ford Mustang Turbo GT/Mercury Capri Turbo RS

Manufacturer: Ford Motor Company
Rotunda Dr.
Dearborn, MI 48121

Model year (s):   1983—(continuing)
Specifications for: 1984 Ford Mustang Turbo GT/Mercury Capri Turbo RS
Vehicle type: front engine, rear drive
Body styles: 3-door sedan

Wheelbase: 100.5 in
Length: 197.1 in.
Width: 67.4/69.1 in
Height: 51.9 in
Weight: 2940 lbs
SAE volume, interior/trunk: 84/12
Fuel tank capacity: 15.4 gal.

Engine Type: 4 in-line turbo gasoline
Displacement: 2295cc/140 cu in
Turbo: AiResearch TO3 with integral wastegate
Maximum boost: 10 psi
Compression ratio: 8.0:1
Fuel system: electronic ported fuel injection
Horsepower: 145 @ 4600 rpm

Transmission (s):   5-speed manual

Front suspension: ind., MacPherson strut, coil springs
Rear suspension: rigid axle, coil springs

Acceleration, 0–60 mph: 8.8 sec.
Top speed: 123 mph
Roadholding: .77g
Braking, 70–0 mph: 201 ft.
EPA Fuel economy, city: 23

FORD MUSTANG SVO

174 horsepower! That's more than double the horsepower of the normally-aspirated 2.3 liter four-cyclinder on which the SVO engine is based. This SVO is based on, but is not the same turbo 2.3 liter found in the '83–'85 Thunderbird and the '84 and later Mustang/Capri, which has 145 horsepower. The SVO has the most powerful four cylinder engine available in a production car.

A turbo 2.3 liter mounted to the rear of the engine compartment and special tires and suspension make the Mustang SVO handle like the V-8 never could.

The SVO turbo is the first American car with an intercooler. The intercooler cools the compressed air after it comes out of the turbocharger, but before it reaches the intake manifold. This cooling improves fuel economy and power by creating a denser charge at a given boost pressure (like a hood scoop or any other cold air induction system) and reducing the engine's tendency to detonate, thus permitting a higher boost pressure to be used. The intercooler looks and works like a radiator, and the special SVO hood scoop forces cold air across its fins. Because the turbocharger naturally heats the air while it compresses it (as predicted by Boyles' law), the temperature of the compressed air can drop as much as 90°F across the intercooler. The SVO also has an engine-mounted oil–to–water oil cooler which uses engine coolant to cool the oil.

The SVO features an extremely sophisticated electronic control system. It is based on the 145 horsepower non-intercooled 2.3 liter turbo's electronic controls, except that it can also modulate boost pressure, and hence uses different software.

The choice of software and boost modulation is literally at the driver's fingertips. A two-position cockpit-mounted switch is marked "regular" and "premium," denoting the recommended positions for low and high octane gasolines. In the "regular" position, the engine operates like the non-intercooled 2.3 liter turbo, with a maximum of 10 psi boost and no electronic-control of the wastegate. The detonation sensor will cause ignition retard, but not boost reduction. The result is only slightly more than the 145 horsepower of the basic 2.3 liter turbo. This slight power improvement is due to the intercooler, and the actual increase depends on ambient air temperature and flow through the intercooler.

But the SVO really comes alive when you fill up with high test and flip the switch to "premium." A maximum of 14 psi boost is now available, and the electronic controls modulate both ignition timing and boost pressure to produce a steady 210 ft/lbs of torque from 3000 rpm to 4400 rpm. This steady flow of torque produces a linear rise in horsepower from 3000 rpm up to 4400 rpm, at which point 175 horsepower is available, assuming no detonation.

The computer's decision on exactly how much to reduce boost pressure and retard ignition timing when detonation is present is a complex procedure. First it must de-

The Mustang SVO's intercooler sits atop the engine. A hood scoop forces air through the fins at speed.

termine throttle position, as expressed by the throttle position sensor. If the position is from closed to one third throttle, one set of strategies is used, from one third to two thirds throttle another set of strategies is used, and from two thirds to full throttle still another set is used. Each set of strategies contains four different modulation maps. The computer must analyze detonation strength, coolant temperature, air flow, and engine speed before selecting a modulation map.

When detonation is detected, the computer immediately retards the ignition timing and signals the wastegate solenoid to reduce boost pressure. Actual intake manifold pressure reduction lags a split second behind ignition retard, due to the time solenoid and wastegate mechanical action require. Ignition retard and boost reduction continues step by step until detonation is eliminated. Ignition timing can be retarded by as much as 20°, and boost pressure can be brought down as low as 10 psi.

In addition, the computer can compare the actual detonation strength with the expected detonation strength. If detonation is not being reduced quickly enough by the modulation map, the computer will order the wastegate solenoid to stay open until the computer is convinced that detonation is almost eliminated.

The computer also controls the electronic port fuel injection, idle speed, and an exhaust gas recirculation on/off solenoid. The fuel injection control computer uses an oxygen sensor, barometric pressure sensor, coolant temperature sensor, throttle position sensor, air conditioning sensor (to indicate increase load for idling), and a vane air flow sensor with an integral air temperature sensor.

The SVO accelerates to 60 mph in 7.4 seconds, with a top speed of 128 mph. EPA fuel economy is expected to be 21 mpg city, 32 mpg highway. Top speed and fuel economy are enhanced by the special aerodynamic front end and rear wheel spats which direct air flow around the wheel wells. The polycarbonate plastic dual-wing rear spoiler doesn't merely reduce lift, it actually creates a downforce on the tires which increases traction in the rear.

Mustang SVO's polycarbonate dual wing rear spoiler actually creates a downforce at speed. At 128 miles per hour, it makes a big difference in stability.

Sophisticated electronic controls handle the Mustang SVO's port fuel injection, boost pressure, ignition timing, and other functions. The net result is 174 horsepower out of a mere 2.3 liters.

It's no surprise that the SVO is much more maneuverable and boasts quicker response to steering inputs than the V-8 Mustang. The reason is a surprise, though. It's not because there's less weight on the front wheels, because there isn't. The SVO turbo 2.3 liter weighs almost as much as the 302 V-8. However, the SVO engine weight is mostly behind the front wheels, because the four-cylinder is smaller and can be mounted well to the rear of the engine compartment. Placing the engine weight closer to the center of the car reduces the polar moment of inertia, that is, the car's resistance to turning.

Early SVOs will use the Goodyear European NCTs. These 50-series tires are 16 inches in diameter and V-rated for sustained speeds well above even the SVO's capability. Seven inch wide aluminum wheels are also standard. Later SVOs will use the new V-rated directional Goodyear Eagle VR50s similar to those used on the '84 Corvette. The SVO has its own heavy-duty spring rates and anti-sway bars. Koni nitrogen shock absorbers and MacPherson struts are adjustable. Steering gain and feedback is improved by the moderately quick ratio (20:1) power rack and pinion steering. At speeds that have the V-8 Mustang 302 GT on the verge of rolling over in the curves, the SVO feels flat, stable, and composed—as if it where wondering what all the 302 GT's commotion is about.

The SVO comes with four wheel disc brakes and Recaro-style seats. The five speed transmission uses a Hurst shift linkage. Brake and gas pedals are relocated to facilitate heel and toe downshifts. A rear window defroster, intermittent windshield wipers, AM/FM stereo radio, leather shift knob and leather tilt steering wheel are all standard on the SVO. The only major options are air conditioning, power windows, power door locks, cassette deck, flip-up sunroof, and leather seat trim.

Model: Ford Mustang SVO

Manufacturer: Ford Motor Co.
                   Rotunda Dr.
                   Dearborn, MI 48121

Model year (s): 1984—(continuing)
Specifications for: 1984 Ford Mustang SVO
Vehicle type: front engine, rear drive
Body styles: 3-door sedan

Wheelbase: 100.5 in
Length: 181.0 in
Width: 69.1 in
Height: 51.9 in
Weight: 2990 lbs
SAE volume, interior/trunk: 84/12
Fuel tank capacity: 15.4 gal.

Engine type: 4 in-line turbo gasoline
Displacement: 2295cc/182.7
Turbo: AiResearch TO3 with electronically-controlled integral wastegate, intercooler
Maximum boost: 14 psi
Compression ratio: 8.0:1
Fuel system: electronic port fuel injection
Horsepower: 174 @ 4400 rpm

Transmission (s): 5-speed manual

Front suspension: ind., MacPherson strut, coil springs
Rear Suspension: rigid axle, coil springs

Acceleration, 0–60 mph: 7.5 sec.
Top speed: 128 mph
Roadholding: .77g
Braking, 70–0 mph: 217 ft
EPA Fuel economy, city/hwy: 21/32

FORD THUNDERBIRD/MERCURY XR-7 TURBO

Though their rear ends look radically different, Ford's Thunderbird Turbo Coupe and Mercury's Cougar XR-7 are sisters under the skin. While the T-bird sports an aerodynamic body in the European idiom, the XR-7 has a severe, Buick Regal-like notchback. But their chassis and drivetrains are identical.

Powering these twins is Ford's new 2.3 liter turbo. Electronic ported fuel injection, detonation sensor, and Ford's EEC IV computerized engine control system are all part of the 145 horsepower-producing package. A tuned intake manifold (with 20.6 inch long runners) increases power at low engine speeds, where turbo boost is minimal. Pistons are forged aluminum and the valve cover is cast aluminum.

Ford Thunderbird Turbo Coupe features aerodynamic sheetmetal, fog lights, quick ratio steering, and good handling.

The Thunderbird Turbo Coupe's sleek lines are the envy of the highway.

T-bird/XR-7 Turbo engine is a 2.3 liter port fuel injected 4-cylinder. An AiReserch turbocharger pumps it up to 145 horsepower.

The new T-bird and XR-7 use the same basic chassis as their immediate predecessors. Nitrogen shocks and struts are new. One problem with the earlier versions was a limply located rear axle. Under acceleration and hard cornering the rear axle would hop and squirm around. To solve this problem two horizontal nitrogen-filled shock absorbers have been added to restrict the axle's motion. Unlike the earlier versions, the new T-bird and XR-7 handle and stop reasonably well. The power rack and pinion steering ratio varies from 15:1 at the center to 13:1 at the stops—a quicker ratio than even the Mustang GT or Capri RS! A limited slip rear axle is standard.

For long distance haulers, the XR-7 comes with a 20.6 gallon gas tank—the sleek

Notchback Mercury XR-7 Turbo has a big 20.6 gallon gas tank, and gets 22 EPA city miles per gallon.

Thunderbird Turbo Coupe only has room for 18.0 gallons. The Turbo Coupe comes with Marchal fog lights in an air dam. Unusual options on both models include a keyless entry system, and a Tripminder Computer which tells the time, date, trip distance and elapsed time, average speed, fuel used, and average and instantaneous fuel economy. A traveler's assistance kit includes a flashlight (batteries included), gloves, screwdrivers, fuses, bulbs, pliers, tire pressure gauge, triangular emergency reflector, first aid kit, and jumper cables.

Model: Ford Thunderbird Turbo Coupe

Manufacturer: Ford Motor Co.
Rotunda Dr.
Dearborn, MI 48121

Model year (s):   1984—(continuing)
Specifications for: 1984 Ford Thunderbird Turbo Coupe
Vehicle type: front engine, rear drive
Body styles: 2-door sedan

Wheelbase: 104.0 in
Length: 197.6 in
Width: 71.1 in
Height: 53.2 in
Weight: 3070 lbs
SAE volume, interior/trunk: 92/15
Fuel tank capacity: 18.0 gal.

Engine type: 4 in-line turbo gasoline
Displacement: 2295cc/140 cu in
Turbo: AiResearch TO3 with integral wastegate
Maximum boost: 10 psi
Compression ratio: 8.0:1
Fuel system: electronic ported fuel injection
Horsepower: 145@ 4600 rpm

Transmission (s):   5-speed manual, 3-speed auto

Front suspension: ind., MacPherson strut, coil springs
Rear suspension: rigid axle, coil springs

Acceleration, 0–60 mph: 9.1 sec
Top speed: 123 mph

Roadholding: .77g
Braking, 70–0 mph: 201 ft
EPA Fuel Economy, city/hwy: 22/33

Model: Mercury Cougar XR-7

Manufacturer: Ford Motor Co.
                Rotunda Dr.
                Dearbrn, MI 48121

Model year (s):   1984—(continuing)
Specifications for: 1984 Mercury Cougar XR-7
Vehicle type: front engine, rear drive
Body styles: 2-door sedan

Wheelbase: 104.0 in
Length: 197.6 in
Width: 71.1 in
Height: 53.4 in
Weight: 3070 lbs
SAE volume, interior/trunk: 92/15
Fuel tank capacity: 20.6 gal.

Engine type: 4 in-line turbo gasoline
Displacement: 2295cc/140 cu in
Turbo: AiResearach TO3 with integral wastegate
Maximum boost: 10 psi
Compression ratio: 8.0:1
Fuel system: electronic ported fuel injection
Horsepower: 145 @ 4600 rpm

Transmission (s):   5-speed manual, 3-speed auto.

Front suspension: ind., MacPherson strut, coil springs
Rear suspension: rigid axle, coil springs

Acceleration, 0–60 mph: 9.2 sec
Top speed: 120 mph
Roadholding: .76g
Braking, 70–0 mph: 201 ft
EPA Fuel economy, city/hwy: 22/31

MITSUBISHI CORDIA TURBO

Mitsubishi's Cordia Turbo is in direct competition with two other lightweight four cylinder four seat front wheel drive performance hatchbacks: The Dodge Charger 2.2 and Volkswagen Scirocco. And the Cordia falls right into the thick of the fray, though the Dodge emerges the performance champ. The Cordia's acceleration is a tad slower than the Charger, but quicker than the Scirocco. The Cordia's roadholding and fuel economy are worse than the competition. The Cordia has slightly more passenger room but less luggage space than the Charger and Scirocco. Price wise, it costs more than the Charger but less than the Scirocco.

The transversely-mounted 1.8 liter engine uses a Mitsubishi TCO5 turbocharger, the second largest turbo Mitsubishi makes, and the largest Mitsubishi uses on a car in the U.S. Eight pounds of boost brings the horsepower up to 116 from the normally-aspirated version's 82 horsepower, reducing 0–60 mph acceleration time by three and a half seconds. EPA city fuel economy is reduced by eight mpg with the turbo. Only a 5-speed transmission is available.

Mitsubishi's front wheel drive performance hatchback is the Cordia Turbo.

Mitsubishi's own turbocharger boosts the 1.8 liter Corida Turbo engine up to 116 horsepower. Twin engine counter-balancers reduce vibration.

The Cordia turbo features a rear anti-sway bar, alloy wheels, 185/70 HR13 raised black letter tires, a black front grille, and a one–spoke steering wheel. Power steering and ventilated front brake rotors are standard. The Turbo sports a front air dam and rear spoiler. The Cordia Turbo's appearance is aggressive. The squared-off rear wheel openings look out of place, but Mitsubishi claims this design reduces air turbulence around the rear wheels.

Model: Mitsubishi Cordia Turbo

Importer: Mitsubishi Motor Sales of America
10540 Tablert St.
Fountain Valley, CA 92708

Model year (s):   1984—(continuing)
Specifications for: 1984 Mitsubishi Cordia Turbo

Vehicle type: front engine, front drive
Body styles: 3-door sedan
Wheelbase: 96.3 in
Length: 173 in
Width: 65.4 in
Height: 49.4 in
Weight: 2380 lbs
SAE volume, interior/trunk: 78/14
Fuel tank capacity: 12.8 gal

Engine type: 4 in-line turbo gasoline
Displacement: 1795cc/109.5
Turbo: Mitsubishi TCO5 with integral wastegate
Maximum boost: 8.0 psi
Compression ratio: 7.5:1
Fuel system: electronic throttle body fuel injection
Horsepower: 116 @ 5500 rpm

Transmission (s): 5-speed manual

Front suspension: ind., MacPherson strut, coil springs
Rear suspensions: ind., coil springs

Acceleration, 0–60 mph: 9.3 sec
Top speed: 106 mph
Roadholding: .75g
EPA Fuel economy, city/hwy: 25/38

MITSUBISHI TREDIA TURBO

Mitsubishi's Tredia Turbo is the least expensive turbocharged four-door sedan you can buy. It combines innocent looks with 9.3 second 0–60 mph acceleration, and offers more interior room than a BMW 318i.

The G62B Turbo engine used in the Tredia Turbo is a 4-cylinder overhead cam design with a displacement of 1795 cc, rated at 116 horsepower at 5500 rpm. It is one of the new generation of engines designed and manufactured by Mitsubishi Motors specifically to be compatible with turbocharging.

For quiet, smooth operation, the engine is equipped with Dual Engine Stabilizers, Mitsubishi Motors' engine counterbalancing system. Mitsubishi says a 4-cylinder engine is inherently rough and noisy because of second-order vibrations in the vertical and rolling directions. These originate from the reciprocating masses and the violent explosions in the cylinders. These vibrations cause increased stress and noise in various portions of the engine and its mountings. At times there are also harmonic effects or "booms" that result from combinations of these vibrations, making their presence even more unpleasant.

In the Mitsubishi engine, counterbalancing shafts are used to dampen both the vertical and rolling components of these second-order vibrations. The key to damping the rolling vibration lies in the placement of the shafts. One is slightly above the uppermost portion of the crankshaft while the other is about halfway up the cylinder block.

The shafts are located in rigid sections of the engine structure to carry the loads that develop during high-speed operation. Each shaft is supported by two bulkheads but all the balance load is taken by the bulkhead in the center of the engine block and no balance or bending loads reach the area of the block surrounding the front bearings.

The upper counterbalancing shaft rotates in the same direction as the crankshaft while the lower turns in the opposite direction. Both shafts are driven at twice crankshaft speed and this is done by a single chain. The lower shaft is driven through a reversing gear which also serves the oil pump.

Mitsubishi Tredia Turbo has more interior room than a BMW 318i, but it's the least expensive turbo four-door sedan you can buy.

The driven sprockets are provided with steel cushioning rings for noise suppression and the reversing gear's oil pump driving function contributes to quiet gear operation. The rear bearings of the shaft are fed oil under pressure from the front bearing through holes in the center of the shaft.

One of the most welcome results of minimized vibration is the reduction of "wows" and "booms" in the passenger compartment in everyday driving. Another result of the reduced vibration is less impact and wear on such engine auxiliaries as emission control equipment, electrical and fuel system components and electronic circuitry for ignition and fuel injection systems.

The turbo engine also uses the MCA-Jet system to improve combustion efficiency. In this design there is a smaller third valve in each cylinder. This is associated with a separate intake port and manifold passage for each cylinder. All the jet valves are fed through a small, unthrottled passage in the throttle body. The system produces a constant turbulence in the combustion chamber, even at idle, when most engines give poor combustion due to an almost complete lack of air swirl. The turbulence is produced because the absence of throttling in the carburetor means air and fuel are always rushing in around the tiny jet valve. The result is more efficient combustion of a superlean mixture which in turn helps to reduce fuel consumption and emissions of carbon monoxide, hydrocarbons, and oxides of nitrogen.

The turbocharger, combined with electronic throttle body fuel injection, increases horsepower 41% over the carbureted normally aspirated version of the engine, up to 116 horsepower from 82. EPA city fuel economy is down a costly 8 mpg from the 1.8 liter normally-aspirated Tredia.

The Mitsubishi TCO5 turbocharger assembly is small and light enough to be held in one hand and complete with wastegate it weighs only 16 lbs. The turbine wheel and compressor are manufactured of high-nickel steel alloy and are capable of withstanding temperatures up to 1200 degrees C. They are precision cast by the lost-wax method, then attached to the turbine shaft by electronic welding. After hand finishing to a tolerance of 1/1000th of a millimeter, they are computer balanced to tolerate rotational speeds of up to 120,000 rpm. Lubrication of the turbine shaft is critical and for this reason the engine is equipped with a high discharge oil pump and an engine oil cooler.

Detonation is controlled by a combintaion of low compression ratio (7.5:1) and detonation sensor-induced ignition retard. The wastegate opens at approximately 8 psi. Maximum boost is achieved at roughly 2000 rpm and there is no perceptible turbo lag after that engine speed.

Other features of the Tredia Turbo include a 5-speed overdrive manual transmission, ventilated front disc brakes, an adjustable column steering wheel, variable ratio power steering, and a specially tuned suspension that includes a rear anti-sway bar.

Model: Mitsubishi Tredia Turbo

Importer: Mitsubishi Motor Sales of America
10540 Tablert St.
Fountain Valley, CA 92708

Model year (s): 1984—(continuing)
Specifications for: 1984 Mitsubishi Tredia Turbo
Vehicle type: front engine, front drive
Body styles: 4-door sedan

Wheelbase: 96.3 in
Length: 172.4 in
Width: 65.4 in
Height: 51.6 in
Weight: 2400 lbs
SAE volume, interior/trunk: 84/11
Fuel tank capacity: 12.8 gal

Engine type: 4 in-line turbo gasoline
Displacement: 1795cc/109.5 cu in
Turbo: Mitsubishi TCO5 with integral wastegate
Maximum boost: 8.0 psi
Compression ratio: 7.5:1
Fuel system: electronic throttle body fuel injection
Horsepower: 116 @ 5500 rpm

Transmission (s): 5-speed manual

Front suspension: ind., MacPherson strut, coil springs
Rear suspension: ind., coil springs

Acceleration, 0–60 mph: 9.3 sec.
Top speed: 102 mph
Roadholding: .74g
Braking, 70–0 mph: 209 ft
EPA Fuel economy, city/hwy: 25/38

MITSUBISHI STARION TURBO

Want to drive around in a high tech sports car most people have never even seen before? New for '83, Mitsubishi's Starion fills the bill with exotic sheetmetal, throttle body fuel injected turbo motor, an interior of the future, fantastic brakes, and very good handling.

The big 2.6 liter four cylinder is pumped up to 145 horsepower, 30 horsepower more than the normally aspirated version used in the Dodge Challenger/Plymouth Sapporo. Mitsubishi makes the tiny turbocharger, which only weighs 16.5 pounds complete, but builds 7.5 psi of boost before its integral wastegate opens. The Mitsubishi ECI fuel injection meters air with a unique ultrasonic air flow meter. There is no flap to impede air flow, as in conventional flow meters. The Mitsubishi unit sits in the air cleaner. A vortex-generating rod creates a minute pulsing of the air flow. An ultrasonic sensor counts the pulses, and relays this information to the computer. On the basis of this air flow info, as well as exhaust oxygen concentration (which gives an indication of the fuel/air mixture, as picked up by an oxygen sensor in the exhaust manifold), atmospheric pressure, air temperature, coolant temperature, engine speed, throttle position, throttle opening or closing speed, and battery voltage, the computer tells the fuel injectors how much gasoline to squirt. Two injectors squirt alternately into a single throttle body for more accurate metering and produce a swirled spray pattern for better fuel dispersion.

Mitsubishi's Starion ES has excellent brakes, a neat interior, and turbo 2.6 liter power.

The 2.6 liter uses a three valve per cylinder MCA Jet cylinder head. In this cylinder head, the rocker arm depresses a second smaller intake valve in step with the normal intake valve, delivering a high speed jet of air into the combustion chamber. This jet flow makes the fuel/air mixture swirl, which accelerates the burning of the mixture.

The Starion's big four cylinder uses two counter–rotating balance shafts to reduce engine vibration. Ignition timing is automatically retarded under boost. The engine also has a detonation sensor on the right side of the engine block, helping the ignition computer (located on the left inner fender) further retard ignition timing by as much as twelve degrees when detonation rears its destructive head.

When equipped with the "Technical-Performance Option Package," the Starion is one of the best–stopping cars on the market, thanks to four wheel ventilated disc brakes and a rear brake anti-lock system. In this anti-lock system, the rotational speed of the rear wheels is constantly monitored via a pulse generator and deceleration is measured by a g-sensor. When the rotation of the rear wheels is slowing down faster than the deceleration of the car, the control unit modulates brake line pressure to the rear wheels, preventing rear wheel lock-up. A limited-slip differential, which improves rear tire traction under acceleration and cornering, is also included in this option package.

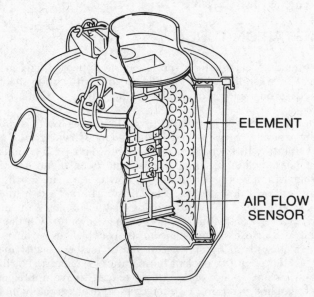

ELEMENT

AIR FLOW
SENSOR

The Starion's air flow sensor sits in the air filter canister.

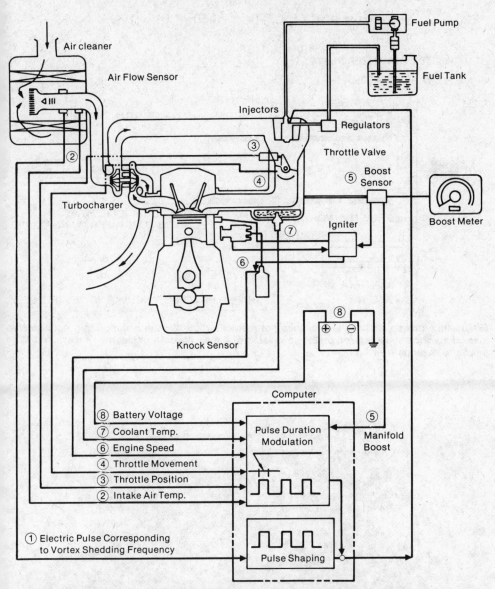

Information concerning battery voltage, coolant and intake air temperature, engine speed, throttle movement and position, air flow, exhaust oxygen concentration, atmospheric and boost pressure is fed to the Starion's computer.

The "Luxury Sport" interior includes six-way adjustable seats: fore-and-aft, backrest and head rest angle, lumbar, side, and thigh adjustments. Leather faced-seats are an option. The LS dash uses LED, FLT, and digital displays. The steering wheel tilts and the headlights go up and down. A neat turbo, Mitsubishi! And for 1984, Dodge and Plymouth are offering the Starion under their own label, the Conquest, with only detail differences.

Function and Operation of Fuel Injection

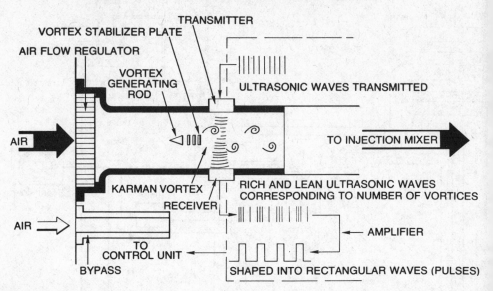

Mitsubishi's unique air flow sensor does not restrict airflow like the common trap door sensor. Instead a vortex shedding rod creates a pulsing of the air flow. An ultrasonic sensor counts the pulses to determine air flow.

Turbocharger is clearly displayed in the Starion's engine compartment. Note the generous heat shielding.

Dodge also sells the Starion—but calls it the Conquest.

Model: Mitsubishi Starion

Importer: Mitsubishi Motor Sales of America
10540 Tablert St.
Fountain Valley, CA 92708

Model year (s):   1983—(continuing)
Specifications for: 1983 Mitsubishi Starion
Vehicle type: front engine, rear drive
Body styles: 3-door sedan

Wheelbase: 95.9 in
Length: 173.2 in
Width: 66.3 in
Height: 51.8 in
Weight: 2820 lbs
SAE volume, interior/trunk: 76/10
Fuel tank capacity: 19.8 gal

Engine type: 4 in-line turbo gasoline
Displacement: 2555cc/155.9 cu in
Turbo: Mitsubishi TCO5 with integral wastegate
Maximum boost: 7.5 psi
Compression ratio: 7.0:1
Fuel system: Mitsubishi ECI fuel injection
Horsepower: 145 @ 5000 rpm

Transmission (s):   5-speed manual

Front suspension: ind., MacPherson strut, coil springs
Rear suspension: ind., MacPherson strut, coil springs

Acceleration, 0–60 mph: 8.7 sec
Top speed: 121 mph
Roadholding: .78g
Braking, 70–0 mph: 182 ft
EPA Fuel economy, city/hwy: 21/31

NISSAN 300ZX TURBO

The Nissan 300ZX Turbo is the fastest car from Japan. It has the most powerful turbocharged engine available in a production car in America, at 200 horsepower. Buick's '84 V-6 turbo shares this power rating, but while the Buick has this much power only in first and second gears, the 300ZX has 200 horsepower available in every gear.

The 300ZX 3.0 liter V-6 engine is new for '84, replacing the rough running but reliable in-line six cylinder used in the 240, 260, and 280Z and 2X models. The V-6 is eight inches shorter and 39 pounds lighter than the in-line six. The V-6 is much smoother than the old in-line, particularly at high engine speeds. Yet the V-6 turbo has a lower redline (6000 vs. 6400 rpm) and produces its maximum power at a lower engine speed (5200 vs. 5600 rpm) than the 280ZX Turbo. The 300ZX Turbo has 20 more horsepower than the 280ZX Turbo, and gets better EPA city gas mileage to boot, up by one mpg to 20. This rating is impressive because only one other 200 or more horsepower car shares such a high EPA city rating: The '84 Porsche 911 Carrera also has 200 horsepower and gets 20 EPA city miles per gallon. None of the other 200 or more horsepower cars (Corvette, V-8 Mustang, Buick V-6 turbo, Porsche 928S, or the exoticars) can muster better than 17 mpg.

To get this excellent combination of power and fuel economy, the 300ZX uses electronic ported fuel injection with Bosch injectors. A Hitachi hot wire flow sensor is used, improving intake flow over the common trapdoor vane air–flow meter. The overhead camshaft cylinder heads are aluminum. They use swirl path intake ports and have a large squish area to reduce detonation. Both normally aspirated and turbo versions use a single main bearing cap casting linking all four bearing caps to strengthen the cast iron block, and the oil pan has reinforcing members welded in.

The turbo version has 7.8:1 compression pistons in place of the normal 9.0:1 pistons. A detonation sensor retards the ignition timing when detonation is present. The Ai-

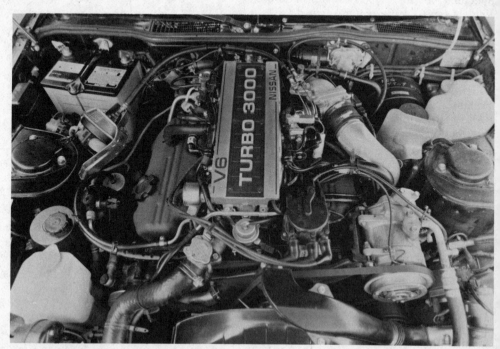

Nissan's Turbo V-6 is the most powerful turbocharged production car engine you can buy at 200 horsepower.

Nissan 300ZX Turbo is fast, sporty, and comfortable.

Research T5 turbocharger begins building boost at a low 1800 rpm, and all 6.7 pounds of boost are available at 2900 rpm.

The 300ZX Turbo gets five lug hubs and bigger 15 x 6 ½″ wheels to handle the turbo's additional power. Handling has traditionally been the Z-car's bugaboo, and while the 300ZX's handling is improved, the jury is still out as to whether the 300ZX is safe to drive really fast.

Model: Nissan 300ZX Turbo

Importer: Nissan Motor Corp.
Box 191
Gardena, CA 90247

Model year (s):   1984—(continuing)
Specifications for: 1984 Nissan 300ZX Turbo
Vehicle type: front engine, rear drive
Body styles: 3-door coupe, 3-door sedan

Wheelbase: 91.3–99.2 in
Length: 170.7–178.5 in
Width: 67.9 in
Height: 51.0–51.6 in
Weight: 3050 lbs
SAE volume, interior/trunk: 52/12
Fuel tank capacity: 19.0 gal

Engine type: V-6 turbo gasoline
Displacement: 2960cc/180.6
Turbo: AiResearch T5 with integral wastegate
Maximum boost: 6.7 psi
Compression ratio: 7.8:1
Fuel system: electronic port fuel injection
Horsepower: 200 @ 5200 rpm

Transmission (s):   5-speed manual, 4-speed auto

Front suspension: ind., MacPherson strut, coil springs
Rear suspension: ind., coil springs

Acceleration, 0–60 mph: 7.1 sec.
Top speed: 135 mph
Roadholding: .78g
Braking, 70–0 mph: 218 ft
EPA Fuel economy, city/hwy: 20/30

NISSAN PULSAR NX TURBO

The Pulsar NX Turbo has the highest EPA city gas mileage of any turbo car you can buy. If fuel economy comes first, but the quicker acceleration of a turbo gasoline engine is important and the sporty but controversial look of the Pulsar appeal to you—it looks like Nissan has caught you. At $8400, the Pulsar NX Turbo is one of the cheapest turbo cars around.

While the normally aspirated Pulsar uses a 1.6 liter engine, the turbo uses a shorter–stroke, 1.5 liter version to keep the additional load from overstressing the block. The overhead cam transversely-mounted four cylinder has been strengthened in other ways, too. The turbo engine has bigger connecting rods, reinforced bearings, heat resistant exhaust valves, and a stronger crankshaft, pistons, rings, and wrist pins. The fuel pump

The Nissan Pulsar NX Turbo offers the best combination of acceleration and fuel economy on the market: 9.7 second 0–60 mph acceleration and 33 EPA city mpg.

Small AiResearch T2 turbocharger and port fuel injection are responsible for the 1.5 liter Pulsar Turbo's 100 horsepower.

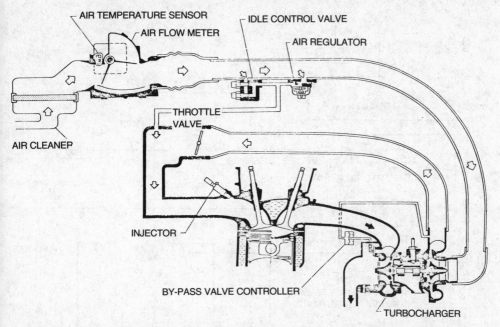

Pulsar's turbocharger sucks air through the air flow meter, and forces the air through the throttle and intake manifold. Port fuel injectors squirt gasoline into the air before the air enters the cylinders.

sits in the gas tank. The compression ratio has been dropped from 9.2:1 to a low 7.8:1. An oil-to-water oil cooler is new. There are two electric fans, one to cool the radiator and one to cool the exhaust system. The exhaust cooler is activated by an exhaust temperature sensor in the exhaust manifold.

The AiResearch turbocharger pumps a maximum of 6.8 psi into the intake manifold before the integral wastegate opens. There is also an emergency relief valve in the intake manifold. Should the boost pressure somehow exceed 7.5 psi, the excess air is bled into the atmosphere.

The electronic port fuel injection works in conjunction with the computerized ECCS engine control system. This microprocessor controls the fuel injection, ignition timing, idle speed, and fuel pump operation. A crank angle/rpm sensor, air flow meter, water temperature sensor, air temperature sensor, barometric pressure sensor, oxygen sensor, and detonation sensor provide the microprocessor with the information it needs to make the big decisions. Automatic transmission cars also have a park/neutral switch and a car speed sensor.

The Pulsar's distributor does double duty as the crank angle/rpm sensor. It works on the same principle as the superb Mallory Unilite, an aftermarket distributor popular among hotrodders. It has a light–emitting diode and two light–sensitive photo diodes. The rotor has a slit for each cylinder. These four slits indicate the crank angle, specifically, the moment when each piston is at the top of its stroke (TDC). Another set of 360 slits, one for each degree of rotation, indicate rpm for ignition timing control. When a slit passes between the LED and a photo diode, the photo diode responds to the light emitted by the LED. This causes a change in voltage and is then converted into an on/off pulse by the wave–forming circuit, which is then sent to the microprocessor.

One photo diode circuit handles the crank angle slits, the other is responsible for the rpm slits. The net result is 100 horsepower, a 43% increase over the normally aspirated version. Yet EPA city fuel economy is only down from 35 mpg to 33 mpg on the turbo. A worthwhile tradeoff, wouldn't you say? Handling is the NX Turbo's weak

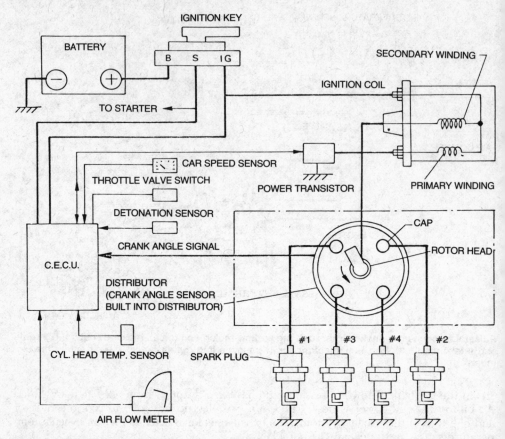

Pulsar's CECU microprocessor takes many factors into consideration before setting ignition timing.

point, despite four wheel independent suspension. The NX Turbo gets slightly stiffer shocks and springs than the normally-aspirated version, but the skinny Toyo radials are so limp, some pick-up trucks can pass the Pulsar around a smooth curve.

Model: Nissan Pulsar NX Turbo

Importer: Nissan Motor Corp.
Box 191
Gardena, CA 90247

Model year (s): 1984—(continuing)
Specifications for: 1984 Nissan Pulsar NX Turbo
Vehicle type: front engine, rear drive
Body styles: 2-door sedan

Wheelbase: 95.1 in
Length: 162.4 in
Width: 63.8 in
Height: 53.3 in
Weight: 2130 lbs
SAE volume, interior/trunk: 80/10
Fuel tank capacity: 13.2 gal

Engine type: 4 in-line turbo gasoline
Displacement: 1488cc/90.8
Turbo: AiResearch T2 with integral wastegate

Maximum boost: 6.8 psi
Compression ratio: 7.8:1
Fuel system: electronic port fuel injection
Horsepower: 100 @ 5200 rpm

Transmission (s):   5-speed manual, 3-speed auto.

Front suspension: ind., MacPherson strut, coil springs
Rear suspension: ind., coil springs

Acceleration, 0–60 mph: 9.7 sec
Top speed: 109 mph
Roadholding: .68g
Braking, 70–0 mph: 202 ft
EPA Fuel economy, city/hwy: 33/46

PONTIAC 2000 SUNBIRD TURBO

Pontiac's 1.8 liter turbo engine puts out more horsepower per liter of displacement than any other production car engine ever. What makes this achievement even more amazing is that the only changes made to the engine block and cylinder head of the normally aspirated version are 8:1 compression ratio pistons, and of course an oil drain-back fitting for the turbo. While the normally aspirated version put out 84 horsepower, the turbo provides 150 horsepower.

It is easy to get confused with J-car engines, because two completely different designs of 1.8 liters have been available. One is a pushrod engine that's no longer available, actually displacing 1841 ccs, with a two-liter big brother. But the 1.8 liter turbo is based on the overhead camshaft 1.8 liter engine. This overhead camshaft engine was originally designed with turbocharging in mind, and that explains the amazing lack of cylinder head and block differences between the normally aspirated and turbo versions.

The similarities end when you get to the manifolds and fuel system. The turbo's intake manifold has tuned runners and is a plenum design. Four port–type fuel injectors are mounted on the intake manifold at the cylinder head flange. The electronic ported fuel injection system uses Bosch's injectors and fuel regulator, but the computer programming is by GM/Delco.

The Delco computer-based detonation control system is very sophisticated, and is

With 150 horsepower and unequal length half-shafts, severe full throttle maneuvers like this create some torque steer in the Pontiac 2000 Turbo's steering wheel.

Power-operated top makes the Pontiac 2000 Turbo convertible fun in the sun.

Pontiac's 1.8 liter port fuel injected turbo has more power per liter than any other production car. Electronically-controlled wastegate and ignition retard system handle detonation.

in good part responsible for the engine's ability to withstand the stress of high power output. The ECM (electronic control module) receives information from an oxygen sensor (which measures the oxygen content in the exhaust for proper regulation of the fuel/air mixture), engine coolant temperature sensor, throttle position sensor, intake manifold air temperature sensor (used to determine air density), vacuum/boost pressure sensor, engine speed sensor, and a detonation sensor. On the basis of this information, the ECM controls the fuel/air ratio, boost pressure, and ignition timing.

Assuming all sensors are reporting normal information and there is no detonation present, the ECM will allow a maximum of 5.2 psi boost at half throttle, 8.2 psi at three-quarters throttle, and 9.4 psi at full throttle. These figures were selected to achieve a linear feeling of power vs. throttle position, and of course there are intermediate steps below and between these figures. The figures given are points on a curve.

Assuming there is no detonation or engine overheating, the ECM will maintain 9.4 psi boost at full throttle forever. However, if detonation is present, the ECM first re-

Pontiac 1.8 turbo uses a tuned runner intake manifold with a plenum design.

tards the ignition timing. After four seconds of detonation, if the ignition timing is still retarded from the norm by at least 8°, the ECM begins to reduce boost pressure at a rate of .45 psi per second. If detonation persists, the ECM can bring boost pressure down to as little as 4 psi. While the ECM is reducing boost pressure it is also still further retarding the ignition timing bit by bit. It can retard the ignition timing by as much as 20° from the norm.

When detonation disappears, boost begins to rebuild only after the ignition timing reaches the point of 4° retard from the norm. When the engine is cold (coolant temperature below 130°F) or overheating (coolant temperature above 234°F), the ECM will only allow 4 psi boost. If any of the sensors are not working properly, a "check engine" light glows on the dash, and boost is limited to 4 psi.

To reduce the chance of torque converter damage caused by abuse, automatic transmission versions will only allow full boost, full throttle converter stall for eight seconds before dropping boost to 4 psi. The ECM also locks up the torque converter in top gear while cruising for improved fuel economy and top speed.

If too much boost pressure builds up in the manifold, due to tampering with the wastegate or component failure, fuel is completely shut off temporarily until boost pressure drops. Similarly, fuel shuts off if the engine is overrevved. Redline is 6500 rpm. If engine speed exceeds 6800 rpm, fuel is shut off until the engine speed comes down below 4500 rpm.

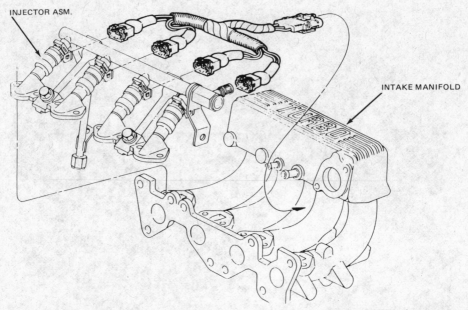

INJECTOR ASM.

INTAKE MANIFOLD

FUEL INJECTOR ASSEMBLY

Four port fuel injectors mount on the Pontiac turbo's intake manifold. Fuel rail provides the gasoline, and the ECM microprocessor orders the injectors to fire.

Stainless steel mini-header feeds the Pontiac's AiResearch turbocharger with exhaust gases.

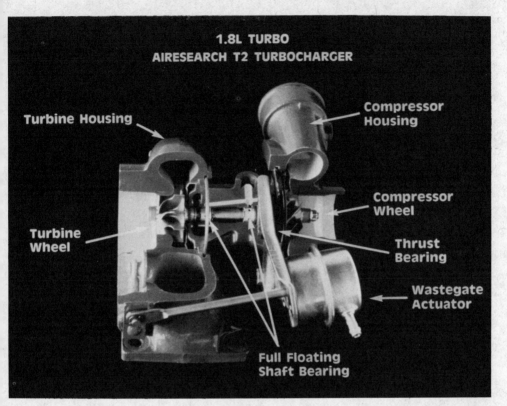

1.8L TURBO
AIRESEARCH T2 TURBOCHARGER

Turbine Housing

Compressor Housing

Turbine Wheel

Compressor Wheel

Thrust Bearing

Wastegate Actuator

Full Floating Shaft Bearing

AiResearch T2 turbocharger is used on the 1.8 liter Pontiac. It is smaller than the common T3.

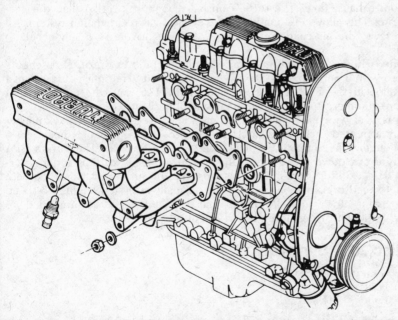

The 1.8 liter long block is built and tested on natural gas in Brazil. The intake manifold and turbo equipment are fitted by Pontiac in the U.S., and then the engine is retested.

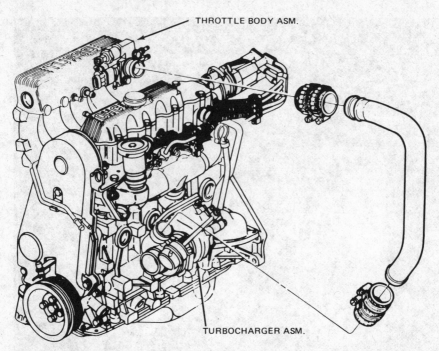

THROTTLE BODY ASM.

TURBOCHARGER ASM.

Crossover tube guides the compressed air from the turbocharger to the throttle body which is mounted on the Pontiac's intake manifold.

An AiResearch T2 turbocharger is used in the Pontiac; this unit is smaller than the common T3. The turbocharger is mounted on a stainless steel tubular exhaust header, Indy-style, rather than the common cast iron exhaust manifold. This tubular design promotes optimum exhaust flow. The turbo camshaft has more overlap than the normally aspirated cam. This allows elimination of the exhaust gas recirculation system. All the accessories except the air conditioner compressor are driven by a single belt to reduce power loss.

The base engine, without manifolds and turbocharger, is built in Brazil. Each engine is hot tested there with a natural gas system before shipment. The intake and exhaust systems are installed at the Pontiac engine plant in the U.S. and the engine is hot tested again before shipment to the vehicle final assembly plant.

The net result is a car which accelerates to 60 mph in 9.1 seconds, 4 seconds quicker than the normally aspirated OHC version. Unfortunately the GM J-car is heavy for a car of its interior dimensions, limiting the Pontiac 2000's performance. The turbo's EPA city fuel economy drops down to 24 mpg with the turbo versus 29 mpg for the normally aspirated OHC 1.8 liter. Part of the reason is that the turbo must make do with GM's wide ratio 4-speed transmission because the normally-aspirated engine's 5-speed isn't strong enough for turbo duty. Premium unleaded gasoline is recommended for maximum power, although 87 octane unleaded can also be used.

Model: Pontiac 2000 Sunbird Turbo

Manufacturer: General Motors Corp.
　　　　　　　　　Pontiac Div.
　　　　　　　　　Pontiac, MI 48053

Model year (s):   1984—(continuing)
Specifications for: 1984 Pontiac 2000 Sunbird Turbo
Vehicle type: front engine, front drive

Body styles: 2-door sedan, 3-door sedan, 4-door sedan, 2-door convertible
Wheelbase: 101.2 in
Length: 175.4 in
Width: 65.9–66.2 in
Height: 51.9–53.8 in
Weight: 2600–2700 lbs
SAE volume, interior/trunk: 84/12 (l2-door), 83/16 (3-door), 91/13 (4-door)
Fuel tank capacity: 13.6 gal

Engine type: 4 in-line turbo gasoline
Displacement: 1796cc/109.5 cu in
Turbo: AiResearch T2 with electronically-controlled integral wastegate
Maximum boost: 9.4 psi
Compression ratio: 8.0:1
Fuel system: Delco electronic port fuel injection with Bosch L-Jetronic injectors
Horsepower: 150 @ 5600 rpm

Transmission (s): 4-speed manual, 3-speed auto.

Front suspension: ind., MacPherson strut, coil springs
Rear suspension: ind., coil springs

Acceleration, 0–60 mph: 9.1 sec
Top speed: 120 mph
Braking, 70–0 mph: 215 ft
EPA Fuel economy, city/hwy: 24/36

PONTIAC FIREBIRD TURBO TRANS AM

This, the world's second production turbocharged V-8 powered car (The Oldsmobile Jetfire F-85 aluminum 215 V-8 turbo was first) was only available for two years: 1980 and 1981. Pontiac hoped it would match the 400 cubic inch T/A in speed, yet get better gas mileage with only 301 cubic inches. But, in comparison with the 1979 Trans Am 400, the turbo was a second and a half slower in 0–60 mph acceleration, and ten miles per hour slower in top speed. EPA city fuel economy was up to 14 mpg, however, two miles per gallon better than the 1979 400 T/A.

An AiResearch TBO305 turbocharger with an integral wastegate pumps the Pontiac 301 up to nine psi for a total of 205 horsepower at 4000 rpm. The '79 T/A 400 had 225 horsepower though the '78 had a meager 200. In the turbo Trans Am, cool air is rammed into a four inch diameter duct over the front spoiler. The turbocharger sucks the fuel/air mixture through a 700 cfm Rochester four-barrel carburetor, pressurizing a special aluminum intake manifold. A water jacket surrounds the intake manifold's plenum chamber, heating the fuel/air mixture for cold driveability and better warm—engine vapori-

Pontiac Trans Am Turbo with V-8 power and the WS 6 handling package make it go as good as it looks.

The Turbo T/A was available in a limited production pace car version in 1980.

The most powerful turbo engine ever found in an American production car, the Pontiac 301 Turbo put out 205 horsepower.

An AiResearch TB0305 turbocharger sucks through a Rochester Quadrajet 4-barrel carburetor on the Pontiac turbo V-8.

zation until the coolant reaches 217°F degrees, at which point the thermostat shuts off flow.

A knock sensor, similar to Buick's, assists the distributor in maintaining optimum ignition timing. More cast iron is added to the normally aspirated 301 block's bearing webs and top deck, and the main bearing cap bolts are one half inch in diameter, instead of the usual seven-sixteenths of an inch. The crankshaft fillets are pressure rolled for improved fatigue resistance. The camshaft has less duration and overlap than in the normally–aspirated 301 to flatten the torque curve.

It almost seems Pontiac didn't want drivers to know the T/A was a turbo. No boost gauge was available. The AiResearch turbo doesn't even whistle while it spins. And the mandatory automatic transmission masks any turbo idiosyncracies. Emissions regulations wrecked chances for a four-speed turbo T/A. Also, the turbo T/A was never available in California, because of that state's stiff smog laws.

The WS6 handling package makes the T/A a fine, smooth road handler, and the four-wheel discs stop the T/A like an anchor. Though the turbo T/A was a quick car, the even faster '79 400 made life rough for the short lived Turbo Trans Am.

Model: Pontiac Firebird Turbo Trans Am

Manufacturer: Pontiac Motor Division
One Pontiac Plaza
Pontiac, MI 48053

Model year (s): 1980–81
Specifications for: 1980 Pontiac Firebird Turbo Trans Am
Vehicle type: front engine, rear drive
Body styles: 2-door sedan

Wheelbase: 108.2 in
Length: 197.1 in
Width: 73.0 in
Height: 49.3 in
Weight: 3600 lbs
Fuel tank capacity: 21 gal

Engine type: V-8 Turbo gasoline
Displacement: 4940cc/301 cu in
Turbo: AiResearch TBO 305 with integral wastegate
Maximum boost: 9 psi
Compression ratio: 7.5:1
Fuel system: One 4bbl Rochester carburetor
Horsepower: 205 @ 4000 rpm

Transmission (s):   3-speed auto

Front suspension: ind., control arms, coil springs
Rear suspension: rigid axle, leaf springs

Acceleration, 0–60 mph: 8.2 sec
Top speed: 116 mph
Roadholding: .81g
Braking, 70–0 mph: 186 ft
EPA Fuel economy, city: 14

PORSCHE 924 TURBO

Do you demand 7.5 second 0–60 mph acceleration and a fast 130 mph top speed, but still want the best fuel economy possible? Back in 1980, the Porsche 924 Turbo was your only choice. And even today only the Mustang SVO can surpass the 924 Turbo's performance and fuel economy boasts. And even then, only by a small amount.

The Turbo is based on the Porsche 924, which was originally designed as a Volkswagen sports car. The engine is designed by Audi, and though it works hard for a normally aspirated two liter, the 924 is too heavy to get good acceleration from only one hundred horsepower. The 924 never sold well in the U.S. because of what was, by Porsche standards mediocre performance.

The Turbo uses the same block and crank assembly as the 924. Pistons are of deep dish design for a lower 7.5:1 compression ratio and maximum surface area (to cool the mixture during compression). The Turbo cylinder head is new, with the spark plugs

Porsche 924 Turbo: fast, comfortable, with decent fuel economy to boot.

The 924 Turbo's Audi-based 2.0 liter four cylinder greets its turbo with a special cylinder head and lower compression pistons. The result is 143 horsepower.

moving to the intake side. Valves and head gasket are also improved to withstand the higher pressure.

The 924 and 924 Turbo both use the transaxle system.

The front-mounted engine is connected with the rear-mounted transmission/differential assembly by a central tube. Due to the higher engine output, the diameter

924 Turbo's low compression dished piston on the left, normally aspirated 924 piston with valve cutouts on the right.

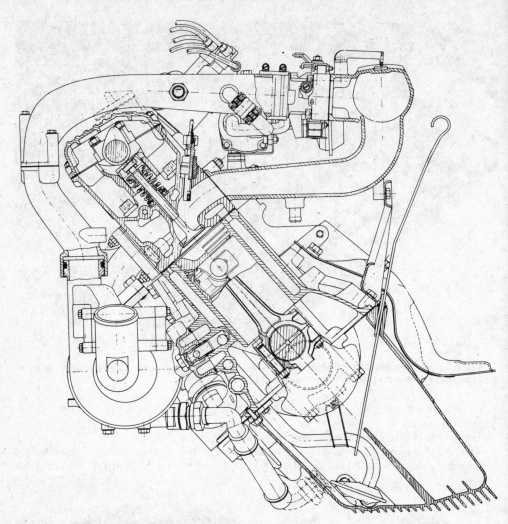

KKK turbocharger blows air through the throttle body which is mounted on the 924 Turbo's intake manifold. Port fuel injectors sit in the cylinder head.

of the 924 Turbo's transaxle drive tube is increased to 25 mm., compared to 20 mm on the normally-aspirated 924.

The NACA duct on the 924 Turbo's hood cools the engine compartment. Slots in the front spoiler direct air to the oil cooler and front brakes.

The optional "Sport Group" package includes 16 inch wheels with wide 55-series Pirelli P7s, stiffer shocks, a rear anti-sway bar, a four-spoke steering wheel, "S" decal, and four wheel disc brakes. The standard rear drum brakes work well in most situations, but repeated hard use will make them fade.

Not to detract from its back road ability, the 924 Turbo is a particularly impressive high speed, long distance highway machine. Its tall overdrive fifth gear, extra sound insulation, and slick aerodynamics keep the cockpit reasonably quiet at three digit speeds. A comfortable interior, fine handling and high speed stability make those hours between fuel stops fun. And the fuel stops can be a long way apart with the 18.6 gallon gas tank.

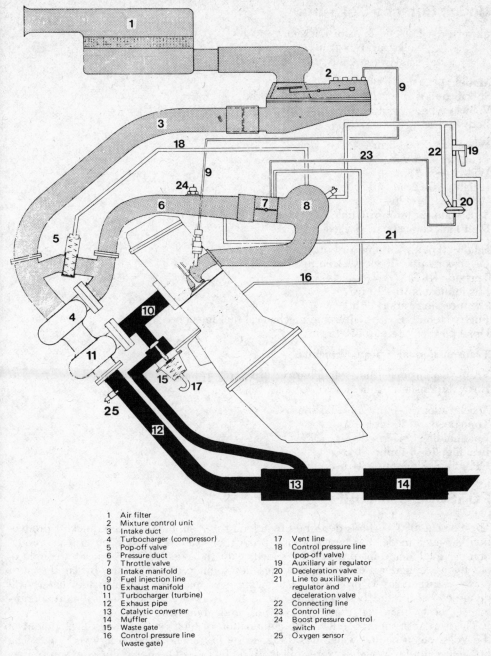

1	Air filter
2	Mixture control unit
3	Intake duct
4	Turbocharger (compressor)
5	Pop-off valve
6	Pressure duct
7	Throttle valve
8	Intake manifold
9	Fuel injection line
10	Exhaust manifold
11	Turbocharger (turbine)
12	Exhaust pipe
13	Catalytic converter
14	Muffler
15	Waste gate
16	Control pressure line (waste gate)
17	Vent line
18	Control pressure line (pop-off valve)
19	Auxiliary air regulator
20	Deceleration valve
21	Line to auxiliary air regulator and deceleration valve
22	Connecting line
23	Control line
24	Boost pressure control switch
25	Oxygen sensor

Layout of the Porsche 924 Turbo's fuel, air, and exhaust system.

In 1982, the 924 Turbo was replaced by the 944. Using a big 2.5 liter normally-aspirated four cylinder, the 944 achieved the same acceleration with better EPA gas mileage. The 944's top speed is lower at 123 mph, however, because it has a two–inch–wider body to clear wider tires.

Model: Porsche 924 Turbo

Importer: Porsche & Audi Division
Volkswagen of America
Englewood Cliffs, NJ 07632

Model year (s):   1980–81
Specifications for: 1981 Porsche 924 Turbo
Vehicle type: front engine, rear drive
Body styles: 3-door sedan

Wheelbase: 94.5 in
Length: 168.9 in
Width: 66.3 in
Height: 50.2 in
Weight: 2700 lbs
SAE volume, interior/trunk: 63/12
Fuel tank capacity: 18.6 gal

Engine type: 4 in-line turbo gasoline
Displacement: 1984cc/121 cu in
Turbo: KKK
Maximum boost: 10 psi
Compression ratio: 7.5:1
Fuel system: Bosch K-Jetronic mechanical fuel injection
Horsepower: 143 @ 5500 rpm

Transmission (s):   5-speed manual

Front suspension: ind., MacPherson strut, coil springs
Rear suspension: ind., torsion bars

Acceleration, 0–60 mph: 7.5 sec
Top speed: 130 mph
Roadholding: .77g
Braking, 70–0 mph: 183 ft
EPA Fuel economy, city/hwy: 20/33

PORSCHE 930 TURBO

The Porsche 930 Turbo is described by a long list of superlatives: the quickest production car ever made; the best stopping production car ever made; the fastest production car ever made in terms of top speed, with the possible exception of a couple of old twelve cylinder Ferraris, if you consider them production cars; by far the most power a turbocharged production car engine ever made; though oversteer haters will disagree, one of the best handling production cars ever made. That's more performance "bests" than you'll find in any other production car ever made.

This superest super beetle is the culmination of high performance development of the Volkswagen beetle, which was designed by Ferdinand Porsche fifty years ago. The first four cylinder Porsches were sold in 1951. In 1964, Dr. Ferry Porsche, son of Ferdinand Porsche, introduced the six cylinder 911. The 930 Turbo, based on the 911 has more than ten times the horsepower of the original beetle, and accelerates to 60 mph in one-eighth of the beetle's time. That's the ultimate story of successful development.

Introduced in 1976, the Turbo Carrera (as it was originally called) was the first turbocharged car to receive EPA emissions certification. The U.S version used two catalytic converters, air injection and exhaust gas recirculation. It has a 234 horsepower three liter version of the famous Porsche flat six engine. The engine had a low, 6.5:1 compression ratio.

The Porsche 930 Turbo is the quickest and best stopping production car ever made. It's also one of the fastest and best handling production cars ever made.

You're staring at a 234 horsepower '76 Porsche Turbo Carrera engine. The '78 version sported an intercooler and 261 horsepower!

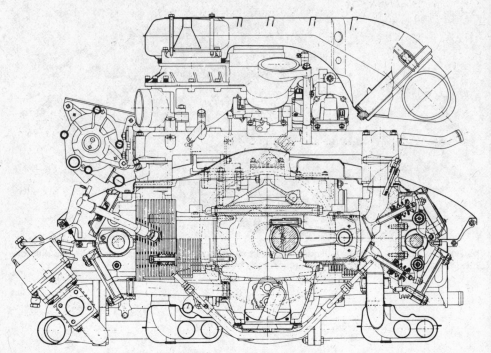

Porsche's infamous turbocharged aircooled flat six cylinder. Nobody else makes anything like it.

The 930 Turbo uses a milder camshaft than the 911. Intake valve lift is 9.7 mm versus 11.6 mm for the 911. Exhaust valve lift is 8.9 mm versus 10.3 mm for the 911. Valve–open periods are also shorter for the Turbo, as is shown below.

	3.0 liter 911	930 Turbo
Intake	24°BTDC–76° ABDC	22°BTDC–62°ABDC
Exhaust	66°BBDC–26°ATDC	52°BBDC–20°ATDC
Overlap	50°	42°

Reducing valve lift reduces inertial forces, making life easier on the valvetrain. Sodium-filled exhaust valves are used to cope with the increased heat. The design of the entire 930 Turbo is highly unique. The rear-mounted engine is aircooled. The six cylinders are horizontally opposed, and the engine crankcases, which split apart like most motorcycles', are magnesium. Each cylinder can be removed from the cases separately. The cylinder heads are aluminum, and each carries its own overhead camshaft.

In 1978, displacement was bumped up to 3.3 liters, compression ratio was increased to 7:1, and an air–to–air intercooler was added. These changes cleaned up the throttle response, and increased horsepower to 261 @ 5500 rpm. 60 mph could be reached in 4.9 seconds, and top speed was 160 mph. Only 350 of these '78s were sold in the U.S. In 1979, tightening emissions standards reduced horsepower slightly, and in 1980 Porsche stopped selling the mighty 930 Turbo in America.

Model: Porsche 930 Turbo

Importer: Porsche & Audi Division
Volkswagen of America
Englewood Cliffs, NJ 07632

Model year (s):   1976–79

Specifications for: 1979 Porsche 930 Turbo
Vehicle type: rear engine, rear drive
Body styles: 2-door sedan

Wheelbase: 89.4 in
Length: 168.9 in
Width: 69.9 in
Height: 51.6 in
Weight: 3040 lbs
SAE volume, interior/trunk: 56/5
Fuel tank capacity: 21.1 gal

Engine type: flat 6 turbo gasoline
Displacement: 3299 cc/201.3 cu in
Turbo: KKK with intercooler
Maximum boost: 11.8 psi
Compression ratio: 7.0:1
Fuel system: Bosch K-Jetronic mechanical fuel injection
Horsepower: 253 @ 6500 rpm

Transmission (s):   4-speed manual

Front suspension: ind., MacPherson strut, torsion bars
Rear suspension: ind., torsion bars

Acceleration, 0–60 mph: 5.4 sec
Top speed: 153 mph
Roadholding?   .81g
Braking, 70–0 mph: 168 ft
EPA Fuel economy, city: 12

RENAULT FUEGO TURBO

One of the least expensive turbocars sports one of the most sophisticated powerplants on the market. Renault's Fuego Turbo has an air-to-air intercooled, Bosch L-Jetronic electronic fuel injected engine. Only three other cars come with an intercooler: the Mustang SVO, Audi Quattro and 5000S. The Langer and Reich intercooler cools the compressed air before it reaches the intake manifold. This cooling improves fuel economy and power by creating a denser charge (like a hood scoop or any other cold air induction system) and reducing the engine's tendency to detonate. The intercooler looks and works like a radiator, and the Fuego Turbo even has an electric fan to increase air flow. Because the turbocharger naturally heats the air well above outside temperature

The least expensive car you can buy with an intercooler, Renault's Fuego Turbo actually gets better EPA gas mileage than its normally aspirated brother.

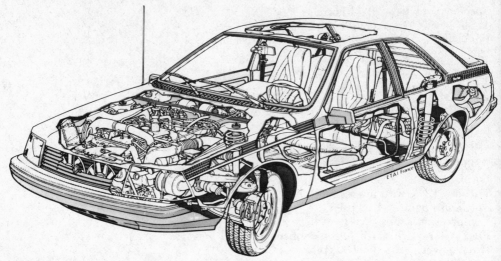

Cutaway of the Renault Fuego Turbo. Note intercooler in the front right corner of the engine compartment.

while it compresses it (according to Boyles' law), the temperature drop of the compressed air is as much as 90°F. There's more information on intercoolers in Chapter Two.

The result is a turbocar which delivers both better acceleration and gas mileage than its normally aspirated version. The Fuego Turbo gets two miles per gallon better EPA city gas mileage than the standard 1.6 liter Fuego! A taller rear axle ratio in the turbo assists in this coup, which also reduces engine rpm on the highway. Unleaded premium is the recommended fuel, though the detonation sensor will retard the ignition timing when necessary for safe use of 87 octane unleaded.

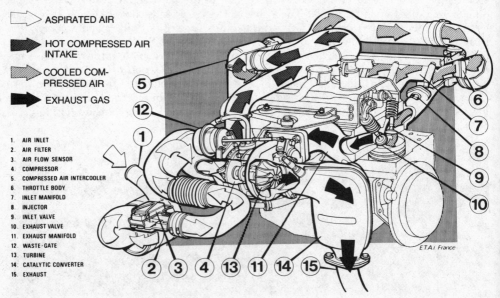

ASPIRATED AIR

HOT COMPRESSED AIR INTAKE

COOLED COMPRESSED AIR

EXHAUST GAS

1. AIR INLET
2. AIR FILTER
3. AIR FLOW SENSOR
4. COMPRESSOR
5. COMPRESSED AIR INTERCOOLER
6. THROTTLE BODY
7. INLET MANIFOLD
8. INJECTOR
9. INLET VALVE
10. EXHAUST VALVE
11. EXHAUST MANIFOLD
12. WASTE-GATE
13. TURBINE
14. CATALYTIC CONVERTER
15. EXHAUST

This illustration makes it easy to trace the movement of air and exhaust gases in the Fuego Turbo engine.

An orfice in the air intake system between the turbocharger and throttle reduces boost to about eight psi at high rpm. This control of boost at maximum rpm is needed because the fuel injection reaches its maximum flow rate under these conditions. At lower engine speeds 12.8 pounds of boost is available.

Also included in the low price is air conditioning, power steering with a tilt steering wheel, a four–speaker stereo, and rear window defogger. Leather seats and an electric sunroof come together as an optional package.

The Fuego features excellent interior room, with enough headroom in the rear for most adults. The ride is smooth. Front and rear anti-sway bars are standard. Handling is limited by the narrow Michelin TRX tires which use special wheels—only TRX tires will fit. To improve roadholding both the wheels and tires must be replaced with high performance units. Be sure to get wheels with the proper offset if you go this route.

Model: Renault Fuego Turbo

Importer: Renault USA
499 Park Ave.
NYC, NY 10022

Model year (s):   1983—(continuing)
Specifications for: 1983 Renault Fuego Turbo
Vehicle type: front engine, front drive
Body styles: 3-door sedan

Wheelbase: 96.1 in
Length: 176.8 in
Width: 66.6 in
Height: 50.5 in
Weight: 2510 lbs
SAE volume, interior/trunk: 83/14
Fuel tank capacity: 14.8 gal

Engine type: 4 in-line turbo gasoline
Displacement: 1565cc/95.5 cu in
Turbo: AiResearch T3 with integral wastegate, intercooler
Maximum boost: 12.8 psi (8 psi at high rpm)
Compression ratio: 8.0:1
Fuel system: Bosch L-Jetronic electronic fuel injection
Horsepower: 107 @ 5500 rpm

Transmission (s):   5-speed manual

Front suspension: ind., control arms, coil springs
Rear suspension: rigid axle, coil springs

Acceleration, 0–60 mph: 10.0 sec
Top speed: 110 mph
Roadholding: .74g
Braking, 70–0 mph: 197 ft
EPA Fuel economy, city/hwy: 26/39

SAAB 900 TURBO

Saab owners consider Saabs a highly original and highly logical answer to the human transportation problem. After a tremendous buildup from this group, I found the car's braking, handling, and fuel economy to be a little nearer to the average than I expected. Thanks to the turbo, this need not be said about the acceleration.

Saab introduced their turbo in 1977. Based on Saab's normally aspirated two-liter

Saab 900 Turbo: a somewhat unusual package, but a good performer.

Saab's turbo 2.0 liter 4 cylinder has an electronically-controlled wastegate to eliminate dangerous detonation. Saab calls this system "APC."

four cylinder, the turbo had a lower compression ratio (7.5:1), different intake valves and camshaft, sodium cooled exhaust valves, an oil cooler, and a larger radiator. It put out 135 horsepower, twenty more than the normally aspirated version, lopping two seconds off 0–60 mph acceleration.

For 1982, Saab engineers developed the APC turbo system to allow the engine to work efficiently with gasolines ranging from 87 to 92 octane. EPA gas mileage is improved by 10% over the non-APC early 1982 Saab Turbo thanks in part to a jump in compression ratio up to 8.5:1. Although maximum horsepower remains unchanged at 135 bhp, torque is up to 172 ft/lbs with 92 octane premium. If you use 87 octane unleaded regular, torque remains the same as with non-APC turbos at 160 ft/lbs.

The key to APC is getting every last ounce of energy out of the gasoline, while avoiding engine-damaging detonation. If low octane gas is used, the engine will detonate more easily. The detonation sensor hears the "knock" of detonation and informs the control unit, which instantly tells a solenoid to open the wastegate for a split-second, to slow the turbo and thereby bleed off intake manifold pressure. This reduced boost pressure eliminates the detonation. The control unit is programmed to tell the solenoid how long to stay open on the basis of information from the detonation sensor, taking engine speed and manifold pressure into account. The APC system can reduce boost to as little as 6 psi.

The Saab's detonation sensor is mounted on the left side of the engine block below the intake manifold. The Boost pressure sensor is mounted on the left inner fender on '82s, and under the instrument panel from '83 on. The control unit sits under the rear seat. The solenoid valve is bolted to the upper radiator crossmember.

Unlike the other turbocharged cars with electronically controlled wastegates, the Pontiac 2000, Mustang SVO, Audi 5000S, and '84 Buick Regal/Riviera, the Saab APC system does not also retard the ignition timing when detonation is picked up by the detonation sensor. The Saab APC system only modulates boost pressure.

Of course Saab owners also love those little Saab extras like heated front seats, filtered high-flow air vents on non-air conditioned models, and cellular bumpers which return to their original shape after a light impact.

Model: Saab 900 Turbo

Importer: Saab Scania of America
Saab Drive
Orange, CT 06477

Model year (s):   1977—(continuing)
Specifications for: 1983 Saab 900 Turbo
Vehicle type: front engine, front drive
Body styles: 3-door sedan, 4-door sedan, 5-door sedan

Wheelbase: 99.1 in
Length: 187.6 in
Width: 66.5 in
Height: 55.9 in
Weight: 2840–2880 lbs
SAE volume, interior/trunk: 88/22-89/14
Fuel tank capacity: 16.6 gal

Engine type: 4 in-line turbo gasoline
Displacement: 1985cc/121.1 cu in
Turbo: AiResearch TCO3 with electronically-controlled integral wastegate
Maximum boost: 9.2 psi
Compression ratio: 8.5:1
Fuel system: Bosch K-Jetronic mechanical fuel injection

Horsepower: 135 @ 4800 rpm

Transmission (s):   5-speed manual, 3-speed auto

Front suspension: ind., control arms, coil springs
Rear suspension: rigid axle, coil springs

Acceleration, 0–60 mph: 9.3 sec
Top speed: 113 mph
Roadholding: .74g
Braking, 70–0 mph: 201 ft
EPA Fuel economy, city/hwy: 21/34

SUBARU TURBO WAGON

On a dry road, the Subaru Turbo Wagon is not terribly exciting. Its braking distances are relatively long. It's slow around curves. Its fuel economy is mediocre. Only a Honda Civic wagon has less interior room, among wagons. And even with the turbo motor, its acceleration is only slightly quicker than average.

But when the pavement ends or the snow begins, the Subaru is in its element. With eight inches of ground clearance and on-demand four wheel drive, the Subaru goes where conventional cars fear to tread. The Subaru's main competition is Toyota's 4WD Tercel, which is far roomier, gets much better EPA fuel economy, stops better, and goes around turns better than the Subaru. But the Subaru has 2.8 inches more ground clearance than the Toyota—plus, the Subaru is quicker accelerating and eight miles per hour faster than the Toyota in top speed. In 0–60 acceleration, the Subaru is a moderate two seconds quicker than the Toyota, but the Subaru gets from 50 to 70 in a whopping 17 seconds less than the Tercel.

The Subaru Turbo is three seconds quicker in 0–60 mph acceleration than the normally-aspirated Subaru wagon yet it gets the same EPA gas mileage as the normally-aspirated automatic. The Turbo has 95 horsepower versus the normally-aspirated engine's 71. L-Jetronic electronic port fuel injection replaces the usual carburetion. The compression ratio has been dropped from 8.7:1 to 7.7:1. A larger radiator, extra fan, and a thermostatically-controlled oil cooler keep the motor cool.

Many owners are oblivious to it, but Subaru's engine layout is highly unique. The cylinders are not in-line, not in a V-formation, but horizontally-opposed, that is, two

At its best when the going gets rough, Subaru's on-demand 4WD Turbo wagon is only available with an automatic transmission.

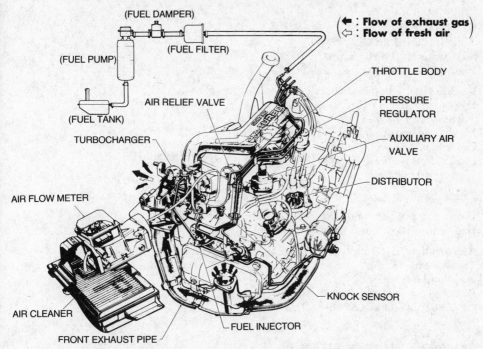

Subaru's unique turbocharged flat 4 cylinder. Even with port fuel injection and a turbo, the powerplant only puts out 95 horsepower. But the Turbo beats the normally aspirated version to 60 mph by 3 seconds, and gets the same EPA gas mileage.

cylinders lie horizontally on each side of the crankshaft. This is similar to the VW beetle, except that the VW was air cooled, while the Subaru is water cooled.

The Subaru Turbo is only available with an automatic transmission. A numerically lower 3.70 axle ratio replaces the usual 3.90 to reduce engine speed and improve fuel economy at highway speeds. First and second gear ratios are lower (higher numerically), and the torque converter's stall speed has been upped from 1900 to 2850 rpm for low speed acceleration. The torque converter locks up in every gear to eliminate wasteful slippage.

Standard equipment with the turbo includes a four-speaker AM/FM stereo, air conditioning, power steering with tilt wheel, and a graphic monitor that indicates brake fluid level, open door, etc. It's no wonder Subaru is the best selling import in Vermont. And for those who like to do their four wheeling in a smaller package, the turbo will be available in the three-door hatchback in mid-1984.

Model: Subaru Turbo Wagon

Importer: Subaru of America
7040 Central Highway
Pennsauken, NJ 08109

Model year (s):   1983—(continuing)
Specifications for: 1883 Subaru Turbo Wagon
Vehicle type: front engine, front drive/4 wheel drive
Body styles: 5-door wagon

Wheelbase: 96.7 in
Length: 168.7 in

Width: 63.8 in
Height: 56.5 in
Weight: 2612 lbs
SAE volume, interior/trunk: 78/29
Fuel tank capacity: 14.5 gal

Engine type: horizontally opposed four turbo gasoline
Displacement: 1782cc/108.7 cu in
Turbo: IHI RHB-52 with integral wastegate
Maximum boost: 7 psi
Compression ratio: 7.7:1
Fuel system: JECS L-Jetronic electronic port fuel injection
Horsepower: 95 @ 4800 rpm

Transmission (s):   3-speed auto.

Front suspension: ind., MacPherson strut, coil springs
Rear suspension: ind., torsion bars

Acceleration, 0–60 mph: 13.7 sec
Top speed: 94 mph
Roadholding: .65g
Braking, 70–0 mph: 224 ft
EPA Fuel economy, city/hwy: 24/30

SUBARU TURBO BRAT

When Subaru dropped their turbo motor into the BRAT, they didn't exactly create a hot rod four wheel driver. But they did greatly improve the BRAT's previously sluggish acceleration and top speed, without reducing the automatic's EPA city fuel economy. 0–60 mph acceleration times dropped a whopping 5.7 seconds, and top speed went up seven miles per hour.

These performance comparisons are between a normally-aspirated five-speed and a turbocharged automatic. The turbo isn't available with a manual transmission, but with performance comparisons like these, that's no great loss. Subaru's automatic lets you switch from front wheel drive to four wheel drive with the press of a button, without

Subaru Turbo BRAT: loads of 4WD fun, but not loads of carrying capacity.

Not as powerful as some turbo cars, but the 4WD Subaru Turbo gets excellent gas mileage.

stopping, even at highway speeds. And the torque converter locks up in every gear for efficient power transfer.

Standard equipment with the Turbo BRAT includes power steering and air conditioning, two features which are optional on most vehicles, and sap horsepower from the engine. A pop-up/removable gull wing T-top, four-speaker stereo, and tilt steering wheel are also standard. The BRAT's ride height can be adjusted up or down with the twist of a few bolts.

The Turbo BRAT is a super–fun four wheel drive with an excellent combination of fuel economy and acceleration, especially for a vehicle with an automatic transmission, power steering and air conditioning. The BRAT falls short on practicality, though, as several domestic and Japanese 4WD pick-ups offer far more load carrying capacity and bed size with similar gas mileage.

Model: Subaru Turbo BRAT

Importer: Subaru of America
7040 Central Highway
Pennsauken, NJ 08109

Model year (s):   1983—(continuing)
Specifications for: 1983 Subaru Turbo BRAT
Vehicle type: front engine, front drive
Body styles: 2-door pick-up

Wheelbase: 96.7 in
Length: 174.2 in
Width: 64.4 in
Height: 56.3 in
Weight: 2400 lbs
Fuel tank capacity: 14.5 gal

Engine type: horizontally opposed four turbo gasoline
Displacement: 1782cc/108.7 cu in
Turbo: IHI RHB-52 with integral wastegate
Maximum boost: 7 psi
Compression ratio: 7.7:1
Fuel system: JECS L-Jetronic electronic port fuel injection
Horsepower: 95 @ 4800 rpm

Transmission (s):   3-speed auto

Front suspension: ind, MacPherson strut, coil springs
Rear suspension: ind, coil springs

Acceleration, 0–60 mph: 13.6 sec
Top speed: 94 mph
Braking, 70–0 mph: 217 ft
EPA Fuel economy, city/hwy: 24/30

VOLVO GLT TURBO

Volvo has long had a reputation for solid, safe automobiles. The GLT Turbo is no exception, even if it is not one of the best packaging jobs on the market, with only 89 cubic feet of interior room for its large 188.8 inch length. By comparison, the Dodge Aries is more than a foot shorter than the Volvo, yet boasts seven cubic feet more interior room.

The Volvo GLT's EPA city fuel economy is 20 mpg. The only sedan with roughly the same interior room as the Volvo that gets worse mileage is the faster Audi 5000 Turbo, which is thirstier by one mpg. The Dodge 600 Turbo gets the same city mileage, but five mpg better EPA highway mileage. The Volvo GLT Turbo's bright spots include its 9.6 second 0 to 60 mph acceleration and powerful four wheel disc brakes. The GLT wagon is the quickest turbocharged station wagon you can buy.

The Volvo's turbo system is conservative. A relatively low five pounds of boost is used for the 2.1 liter four cylinder, and the compression ratio has been dropped way

The Volvo GLT Turbo sedan.

Surprisingly, the Volvo GLT Turbo wagon is a fraction of an inch shorter than the GLT turbo sedan. It's the quickest turbocharged station wagon you can buy.

Using a very conservative turbo system, the Volvo uses only 5 pounds of boost pressure and a 7.5:1 compression ratio. The power increase over the normally aspirated version is only 20 horsepower, for a turbo total of 127.

An AiResearch TB0 3 turbocharger blows through the Bosch K-Jetronic mechanical fuel injection system which is mounted on the Volvo's intake manifold.

down to 7.5:1 from 9.3:1 to prevent detonation. The valves are sodium-filled to dissipate heat, and the valve faces and seats are Stellite, a heat-resistant alloy. A thermostatically-controlled oil cooler keeps the oil temperature safe. The AiResaerch TB03 turbocharger sucks air through a Bosch K-Jetronic air flow sensor. Horsepower is up to 127 from a normally-aspirated 107 horsepower. 0 to 60 mph acceleration is slashed by 3.3 seconds with the turbo, and top speed is up nine miles per hour.

Model: Volvo GLT Turbo Sedan

Importer: Volvo of America
Rockleigh, NJ 07647

Model year (s):   1983—(continuing)
Specifications for: 1983 Volvo GLT Turbo Sedan
Vehicle type: front engine, rear drive
Body styles: 2-door sedan, 4-door sedan

Wheelbase: 104.3 in
Length: 188.8 in
Width: 67.3 in
Height: 56.2 in
Weight: 3020 lbs
SAE volume, interior/trunk: 89/14
Fuel tank capacity: 15.8 gal

Engine type: 4 in-line turbo gasoline
Displacement: 2130cc/130 cu in
Turbo: AiResearch TB03 with integral wastegate
Maximum boost: 5 psi
Compression ratio: 7.5:1
Fuel system: Bosch K-Jetronic mechanical fuel injection
Horsepower: 127 @ 5400 rpm

Transmission (s):   4-speed manual with electrically-operated overdrive, 4-speed auto

Front suspension: ind, MacPherson strut, coil springs
Rear suspension: rigid axle, coil springs

Acceleration, 0–60 mph: 9.6 sec
Top speed: 106 mph
Roadholding: .74g
Braking, 70–0 mph: 196 ft
EPA Fuel economy, city/hwy: 20/29

Model: Volvo GLT Turbo Wagon

Importer: Volvo of America
            Rockleigh, NJ 07647

Model year (s):   1983—(continuing)
Specifications for: 1983 Volvo GLT Turbo Wagon
Vehicle type: front engine, rear drive
Body styles: 5-door wagon

Wheelbase: 104.3 in
Length: 188.5 in
Width: 67.3 in
Height: 57.5 in
Weight: 3200 lbs
SAE volume, interior/trunk: 89/41
Fuel tank capacity: 15.8 gal

Engine type: 4 in-line turbo gasoline
Displacement: 2130cc/130 cu in
Turbo: AiResearch TBO3 with integral wastegate
Maximum boost: 5 psi
Compression ratio: 7.5:1
Fuel system: Bosch K-Jetronic mechanical fuel injection
Horsepower: 127 @ 5400 rpm

Transmission (s):   4-speed manual with electrically-operated overdrive, 4-speed auto

Front suspension: ind, MacPherson strut, coil springs
Rear suspension: rigid axle, coil springs

Acceleration, 0–60 mph: 9.8 sec
Top speed: 106 mph
Roadholding: .74g
Braking, 70–0 mph: 196 ft
EPA Fuel economy, city/hwy: 20/29

4
Turbo-Diesels Factory-Direct

Everybody knows normally aspirated diesels tend to be slow but get great fuel mileage. The promise of the turbo-diesel is a car with good (or at least average) acceleration by gasoline car standards, and great diesel mileage. Some turbo diesels have succeeded in this goal. Others are dismal, wasteful failures: this chapter reveals the major strengths and weaknesses of every turbo-diesel car available in the United States.

When considering the advantages of a turbo-diesel, keep in mind a couple of disadvantages. First, turbo diesels usually cost at least $1000 more than their gasoline counterparts. The fuel savings typically take 80,000 miles to offset this difference in initial cost. Meanwhile, turbo-diesels require oil changes two or three times as frequently as gasoline engines, with filter changes every time, as opposed to every other

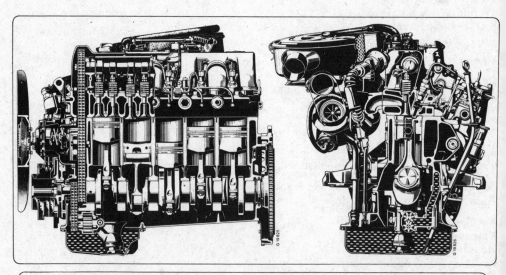

1982 MERCEDES-BENZ 300D/300CD

Five-Cylinder Turbocharged Diesel

Mercedes' 120 horsepower five-cylinder turbo is the most powerful diesel you can get in a passenger car. Typical of Mercedes development, it has been tested and engineered almost to the point of absurdity.

change, as recommended for most gasoline cars. In some cases that means six times as many filter changes as a gasoline engine.

While diesel fuel is slightly less expensive than regular gasoline on the highway, around town diesel usually costs more than regular and about the same as unleaded gasoline. Diesel fuel is less often available than gasoline, so a lack of foresight about refueling is more likely to cause tense moments Sunday night in Nowheresville.

Some diesel owners use home heating oil or kerosene, similar in terms of specific gravity (the weight or density of the fuel as compared to water) to #2 and #1 diesel fuel, respectively. While these fuels in some cases may perform fairly well, these owners are gambling. Diesel engines require fuels meeting a number of different standards, the most important of which is cetane rating. Cetane refers to ignitibility—the ease and speed with which the fuel begins burning after injection starts. Not only will low cetane fuels be less likely to give adequate starting performance in cold weather, they may damage the engine through excessive knock, especially when the engine is cold.

Other parameters such as ash and sulphur content or even fuel cleanliness are important to an engine, but relatively unimportant to a furnace or small kerosene stove or lamp. So don't assume that just because these fuels are basically similar to diesel they will do the job in your car.

Though diesels have a heavy-duty "truck" image, they are actually most efficient under light loads and at low speeds, relative to gas engine performance. While the "easy

Manufactured by Nissan in Japan, the Scout's six cylinder turbo diesel put out 101 horsepower, and was good for 20 EPA city miles per gallon in the Scout II. No doubt a few 1980 model owners are still covered by the 100,000 mile/5 year internal engine warranty!

does-it" EPA estimates are impressive, if you drive fast (especially over 80 miles per hour) or carry a heavy load (especially a trailer), the diesel's fuel efficiency advantage will diminish and may even disappear altogether.

AUDI 5000S TURBO DIESEL

Audi's five-cylinder turbocharged diesel engine is derived from their five cylinder gasoline powerplant. The diesel's block is underbored to thicken and strengthen the cylinder walls for diesel duty. Displacement is hence reduced from 2140 to 1986 ccs. The diesel combustion process reduces power potential, but the turbo helps increase available air and power. The net result is a 16% drop in horsepower in comparison with the normally-aspirated gasoline version, but a jump in EPA city fuel economy from 17 to 28 mpg. Turbo diesel fans will rejoice in the turbo's multi-faceted improvement over the normally-aspirated Audi diesel: horsepower is up 25%, while EPA city fuel economy is actually improved from 26 to 28 mpg.

The 5000S Turbo Diesel is only available with a three-speed automatic transmission, which I found slow to downshift when you put your foot to the floor. An "E" ("Economy") position for the floor-mounted shifter allows the engine to freewheel when you take your foot off the throttle pedal, reducing fuel-consuming drag on the engine at idle or for coasting. The EPA used this "E" mode for their fuel economy tests. A disadvantage of the "E" mode is increased brake usage in most situations.

Compare the Dodge Aries to Audi's turbo diesel. The Dodge has six cubic feet more interior room, is five seconds quicker to 60 miles per hour and eight miles per hour faster in top speed. The Dodge stops better, and goes around a smooth curve faster. The Dodge gets better EPA highway fuel mileage (the same EPA city mileage). The Dodge costs $11,000 less—that's less than half the cost of the Audi.

While the Audi's construction is of the highest quality, the turbo diesel's performance is poor. Power is lacking. Fuel economy is good, but not great. Space efficiency is poor relative to the Dodge. The Aries, which has much more interior room is fully a foot shorter, and an inch narrower. It's also 600 pounds lighter. But, the Audi's braking and roadholding are far better.

The normally-aspirated Oldsmobile Cutlass Ciera diesel gets the same EPA city fuel economy and eight more mpg on the highway than the Audi. The Olds is quicker, stops much shorter, has eight cubic feet more interior room in a half inch shorter body, offers far superior passive safety according to NHTSA crash tests, and costs $8000 less. The Audi's performance numbers don't stack up too well in comparison to many more modern designs. A decision to buy this car would probably depend on subjective matters—its appearance, road feel, sound, etc.

Audi's 5000S Turbo Diesel.

Model: Audi 5000S Turbo Diesel

Importer: Porsche and Audi Division
            Volkswagen of America
            Englewood Cliffs, NJ 07632

Model year (s):   1983—(continuing)
Specifications for: 1983 Audi 5000S Turbo Diesel
Vehicle type: front engine, front drive
Body styles: 4-door sedan

Wheelbase: 105.5 in
Length: 188.9 in
Width: 69.6 in
Height: 54.7 in
Weight: 3060 lbs
SAE volume, interior/trunk: 90/15
Fuel tank capacity: 19.8 gal

Engine type: 5 in-line turbo diesel
Displacement: 1986cc/121 cu in
Turbo: KKK
Maximum boost: 10.3 psi
Compression ratio: 23.0:1
Fuel system: Bosch VE diesel injection
Horsepower: 84 @ 4500 rpm

Transmission (s):   3-speed auto

Front suspension: ind, MacPherson strut, coil springs
Rear Suspension: rigid axle, coil springs

Acceleration, 0–60 mph: 15.9 sec
Top speed: 92 mph
Roadholding: .72g
Braking, 70–0 mph: 213 ft
EPA Fuel economy, city/hwy: 28/36

INTERNATIONAL SCOUT TURBO DIESEL

International Harvester introduced the Scout Turbo Diesel in 1980—at the time the only turbocharged diesel four wheel drive vehicle available in America—not including heavy trucks, of course. But Scout production screeched to a halt for good in October 1980, so only 3082 Scout II turbos, 5000 two wheel drive Traveller turbos, 1467 4WD Traveller turbos, and 832 Terra pick-up turbos were ever built. If you've got one, its a rare bird—hold on to it!

The engine, designated the 6-33T diesel manufactured by Nissan in Japan, is a 198 cubic inch (3.3 liter) in-line six cylinder with a 22:1 compression ratio. It delivers 101 horsepower at 3800 rpm, twenty horsepower more than the normally aspirated version which was available in the Scout from 1976 to 1979.

An AiResearch T0-3 turbocharger provides up to 6.5 psi of boost before the waste-gate opens. Unlike the normally aspirated version, the turbo has an oil jet in each cylinder to cool the pistons, and three special oil passages connect the left and right oil galleries to improve oil flow. A redesigned oil pump increases oil flow from 11.1 gallons per minute up to 12.7 gallons per minute and the oil cooler is enlarged from three cores up to five cores. Redesigned crankshaft throws, bigger camshaft bearing journals, and a stronger oil pump drive spindle assure reliability. The top piston rings use a full keystone design and a ring carrier is added to improve the durability of the

Though noisy and a little rough around the edges, the International Scout II Turbo Diesel is a "one of a kind" four-wheeler.

The Scout Traveller has an eighteen inch longer wheelbase and overall length than the Scout II. Slow, but one unique way to carry a load or the family up a mountain.

Only 832 Terra Turbo pick-ups were ever made. If you've got one, it's already a collectors item—hold on to it!

ring groove. Bigger cylinder head ports and a larger air cleaner element improve engine breathing. Drivetrain choices included a close ratio or wide ratio four-speed manual with 3.54 or 3.73 axle ratios.

The Scout Turbo Diesel got 20 mpg EPA city. It was noisy and slow (0–60 in 18.2 seconds). Top speed was only 83 mph. Maintenance was substantial, with 2500 mile oil change intervals and valve adjustments. But if you've got one, it's a collector's item—so maintain it well. Undoubtedly a few original owners are still protected by their 100,000 mile or 5 year (which ever comes first) internal engine warranty!

Model: International Scout II, Scout Traveler, Scout Terra

Manufacturer: International Harvester
           401 North Michigan Avenue
           Chicago, IL 60611

Model year (s):   1980
Specifications for: 1980 International Scout II, Scout Traveler, Scout Terra
Vehicle type: front engine, rear drive/4 wheel drive
Body styles: 2-door truck, 2-door pickup

Wheelbase: 100–118 in
Length: 166.2–184.2 in
Width: 70.0 in
Height: 66.0 in
Weight: 3831–4131 lbs

Engine type: 6 in-line turbo diesel
Displacement: 3250cc/198 cu in
Turbo: AiResearch T03 with integral wastegate
Maximum boost: 6.5 psi
Compression ratio: 22:1
Fuel system: diesel injection
Horsepower: 101 @ 3800 rpm

Transmission (s):   close ratio/wide ratio 4-speed manual

Front suspension: rigid axle, leaf springs
Rear Suspension: rigid axle, leaf springs

Acceleration, 0–60 mph: 18.2 sec
Top speed: 83 mph
EPA Fuel economy, city/hwy: 20/24

LINCOLN MARK VII TURBO DIESEL

The Lincoln Mark VII, all new in '84, is significantly downsized from the Mark VI. The Mark VII is 13 inches shorter and 350 pounds lighter than the old Mark VI. Unfortunately, the new Mark has lost a gigantic ten cubic feet of interior room and seven cubic feet of trunk space.

Nonetheless, the Mark VII is still a roomy luxury car, and far more fun to drive than the huge Mark VI. Part of the improvement comes from downsizing, and part comes from the new automatically-leveling air suspension. An onboard computer continuously monitors the compressor and air distribution network to each Goodyear rubber air spring. Whenever side-to-side or front-to-rear loads change, the system evaluates the change against programmed standards and directs appropriate fill or vent commands to the air springs. This procedure constantly adjusts spring pressure during braking, acceleration, and cornering. Nitrogen MacPherson struts and shocks are used, as are front and rear sway bars. Four wheel disc brakes and alloy wheels are standard.

The LSC version has thicker sway bars, 40% more dampening, and a computer reprogrammed for 70% stiffer springing up front, 55% stiffer in the rear—for super luxury car handling. The LSC also gets a quicker power steering ratio (15:1 vs the standard 20.4–18.6:1 variable ratio), wider alloy wheels with 65 series tires, fog lights, and a leather interior.

Standard on all Mark VIIs is a digital dash with trip computer, power windows, mirrors, seats, and locks. Air conditioning with a/c and heating ducts for the rear passengers is standard. The Mark VII is the first U.S. car to use the fully integrated, flush-mounted headlights that are so popular in Europe and contribute to an aerodynamic front end. These lights use the replaceable H4 halogen bulbs. Those who cherish ultimate lighting power can replace these bulbs with 80/100 watt Superbulbs.

Thanks to a 25% reduction in aerodynamic drag, a 350 pound weight loss, and a turbo diesel engine, the Mark VII gets 25 EPA city miles per gallon, up 8 mpg from last year's Mark VI. Acceleration is average by gasoline-powered car standards. The 115 horsepower turbo diesel is made by BMW in Germany. It has an aluminum cross-flow cylinder head and an altitude-compensating fuel injection pump. The cylinders are tilted 20° for hood clearance in the Mark VII. The AiResearch T3 turbocharger has an integral wastegate which limits boost at 11.6 psi—that's more boost pressure than any other production car turbo diesel! Compression ratio is 22:1. The glow plugs continue to glow until the engine warms up, reducing cold-start white smoke and improv-

More features and luxury than even a Mercedes! Automatic leveling air suspension, aerodynamic sheetmetal, and massive downsizing makes the Lincoln Mark III an up-to-date luxury car.

BMW makes the six-cylinder turbo diesel in Germany. Note the double oil sump needed to clearance the oil pan over the front crossmember, keeping the engine as low as possible to maintain the low hoodline. The engine is tilted 20° for the same reason.

ing cold driveability. The German-built ZF 4-speed automatic overdrive transmission has a lock-up torque converter in fourth gear.

The Lincoln Mark VII is definitely world-class, offering a driver's combination of luxury, features, and suspension, with good (by luxury car standards) fuel economy. Lincoln did not take the easy way out and tack on a bunch of options.

Model: Lincoln Mark VII Turbo Diesel

Manufacturer: Ford Motor Co.
Rotund Drive
Dearborn, MI 48121

Model year (s): 1984—(continuing)
Specifications for: 1984 Lincoln Mark VII Turbo Diesel
Vehicle type: front engine, rear drive
Body styles: 2-door sedan

Wheelbase: 108.6 in
Length: 202.8 in

Width: 70.2 in
Height: 54.0 in
Weight: 3800 lbs
SAE volume, interior/trunk: 97/15
Fuel tank capacity: 22.3

Engine type: six in-line turbo diesel
Displacement: 2443cc/149 cu in
Turbo: AiResearch T3 with integral wastegate
Maximum boost: 11.6 psi
Compression ratio: 22.0:1
Fuel system: diesel injection
Horsepower: 115 @ 4400 rpm

Transmission (s):   4-speed auto

Front suspension: ind, MacPherson strut, computer-controlled air springs
Rear Suspension: rigid axle, computer-controlled air springs

Acceleration, 0–60 mph: 15.0 sec
Top speed: 105 mph
EPA Fuel economy, city/hwy: 25/39

MERCEDES-BENZ 300SD

Mercedes' five cylinder turbo diesel is the most powerful passenger car diesel on the market, sporting 120 horsepower. It was introduced in the 300SD in July, 1978 in 110 horsepower form. In comparison with the normally aspirated version of the five cylinder which it replaced, the turbo motor increased horsepower by 43%, while weighing only 7% more. EPA city fuel economy increased by 1 mpg with the original turbo engine, and has improved by another three mpg to an excellent 27 mpg with recent engine, drivetrain, and weight updates.

In 1980, a higher lift camshaft brought horsepower up to 120. In 1981, the torque converter was revised with a 1950 rpm stall speed.

Mercedes-Benz's biggest turbo diesel, the 300SD offers a good combination of interior room, fuel economy, and acceleration—but the price is high.

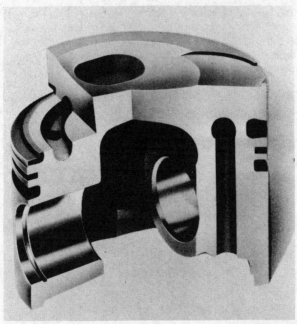

The amount of engineering that has gone into this piston blisters the mind. It has integrated oil-filled cooling pipes, internal and external contours designed to counter heat distortion, external oil chamfers, graphite coating, an alfin-bonded ring support and more.

Many changes were made to the five cylinder while developing it for turbocharging. The crankshaft is now bath-nitrided to increase hardness and fatigue strength. The axial thrust bearings were replaced by bearings with separate thrust plates, and new connecting rod bearings are used.

The pistons received the most intense development. From the start of development fixed oil jets sprayed oil at the interior of the pistons for cooling, but piston temperatures were still as much as 140°F hotter than those in the normally aspirated engine. Pistons with integrated cooling channels just beneath the combustion chamber finally brought temperatures down to levels encountered in the normally aspirated engine. These cooling channels are in the form of a ring, and are cast by inserting a salt core into the mold. After casting, the salt core is melted out.

The internal contour of the piston was designed to minimize distortion. The top land of the piston is contoured to match the heat distortion of the cylinder and to reduce heat flow into the first ring groove. Below the oil scraper ring, the piston is chamfered to spread the oil evenly. The piston skirt is oval, and graphite coated. The piston rings are plasma-sprayed with a molybdenum coating to reduce friction, and an integral ring support for the upper piston ring is alfin-bonded to the piston. The wrist pin bearing is two millimeters larger in diameter, and its axial length in the piston is increased by four millimeters.

The edges of the valve heads were thickened to reduce valve seat wear by improving valve rigidity. The exhaust valves are sodium-filled for cooling. The cylinder head gasket is improved, and the cylinder head is slightly modified to improve heat dissipation in the critical lower area of the prechamber insert. The lubrication system is completely redesigned, using a larger flow chain driven oil pump. Also, the fuel injection period is lengthened slightly to reduce peak pressures and temperatures in the combustion chamber.

The 300SD is the largest Mercedes available with the five cylinder turbo diesel, and is huge, at 202.6 inches overall. Though not space efficient in terms of exterior size, its combination of interior room, fuel economy, and acceleration is outstanding. The engine vibrates slightly at idle and clatters politely during low speed acceleration, but once up to cruising speed, Mercedes' turbo diesel is quieter than most gasoline cars. Oil changes are necessary every 4000 miles, but other than that, maintenance and durability are superb. And Mercedes ride, comfort, and luxury is tops. That's why Mercedes-Benz' are king of the autobahns.

Model: Mercedes-Benz 300SD

Importer: Mercedes-Benz of North America
One Mercedes Drive
Montvale, NJ 07645

Model year (s):   1978—(continuing)
Specifications for: 1983 Mercedes-Benz 300SD
Vehicle type: front engine, rear drive
Body styles: 4-door sedan

Wheelbase: 115.6 in
Length: 202.6 in
Width: 71.7 in
Height: 56.3 in
Weight: 3780 lbs
SAE volume, interior/trunk: 94/15
Fuel tank capacity: 20.3 gal

Engine type: 5 in-line turbo diesel
Displacement: 2998cc/182.9 cu in
Turbo: AiResearch with integral wastegate
Maximum boost: 11 psi
Compression ratio: 21.5:1
Fuel system: Bosch diesel injection
Horsepower: 120 @ 4300 rpm

Transmission (s):   4-speed auto

Front suspension: ind, control arm, coil springs
Rear Suspension: ind, coil springs

Acceleration, 0–60 mph: 13.0 sec
Top speed: 107 mph
Braking, 70–0 mph: 208 ft
EPA Fuel economy, city/hwy: 27/33

MERCEDES BENZ 300TD

There are only two turbo diesel station wagons on the U.S. market: the Volkswagen and the Mercedes. Both are German and both boast very high quality. The Volkswagen gets far better fuel mileage (41 versus 27 EPA city) and only costs one-third as much as the Mercedes. The Mercedes has more interior room and is much quicker than the Volkswagen.

This makes selection of a turbo diesel wagon rather clear-cut. If you don't want to spend megabucks and want the best mileage possible in a station wagon, buy the Volkswagen. If you do have the bucks for a Mercedes and want the extra acceleration, buy the Mercedes. Of course, you could also ask: why buy a diesel at all? A half dozen domestic gasoline-powered station wagons compare favorably with the Mercedes in terms

The Mercedes 300TD is not a terribly cost-effective answer to the fuel economy blues. But it does haul five people and a lot of luggage with teutonic grace.

of interior room, fuel economy, and acceleration. Actually, the Mercedes has the best combination of these three factors. But when a domestic wagon has more interior room and acceleration equal to that of the Mercedes, and only costs $2600 more for fuel in 100,000 miles, can you justify spending $20,000 more for a Mercedes-Benz? Add to that roughly $30,000 saved on loan interest over four years (or gained in your bank account), and you'll hardly notice that the gasoline engine also saved you $550 in oil changes in 100,000 miles. But, there is definitely a market for a station wagon with the Mercedes' excellent combination of fuel economy, acceleration, and interior room.

The 300TD's rear cargo areas has several neat features. The rear seat folds down 60/40 in two sections, so a passenger can sit in the rear, while the other section of the seat folds down to carry more cargo. A containment net reels out of a transverse tube behind the rear seat to keep cargo from flying into the passenger compartment, without hindering rear vision. A color-keyed vinyl sheet pays out of this same tube to hide the freight from prying eyes. The wagon's rear floor panel is a huge trap door, with extra storage room below. An automatic leveling system adjusts the rear suspension's spring preload to match the load. All of which makes for a superb wagon at an unfortunately astronomical price.

Model: Mercedes-Benz 300TD

Importer: Mercedes-Benz of North America
One Mercedes Drive
Montvale, NJ 07645

Model year (s):   1981—(continuing)
Specifications for: 1983 Mercedes-Benz 300TD
Vehicle type: front engine, rear drive
Body styles: 5-door wagon

Wheelbase: 110 in
Length: 190.9 in
Width: 70.3 in
Height: 56.6 in
Weight: 3580 lbs
SAE volume, interior/trunk: 94/41
Fuel tank capacity: 18.5 gal

Engine type: 5 in-line turbo diesel
Displacement: 2998cc/182.9 cu in

Turbo: AiResearch with integral wastegate
Maximum boost: 11 psi
Compression ratio: 21.5:1
Fuel system: Bosch diesel injection
Horsepower: 120 @ 4300 rpm

Transmission (s):   4-speed auto

Front suspension: ind, control arm, coil springs
Rear Suspension: ind, coil springs

Acceleration, 0–60 mph: 13.2 sec
Top speed: 102 mph
Braking, 70–0 mph: 214 ft
EPA Fuel economy, city/hwy: 27/33

MERCEDES-BENZ 300D/300CD

Want to go diesel for fuel economy and durability—but without waiting an eternity to get up to speed? Mercedes' 300D and 300CD are the quickest diesels you can buy. They accelerate from a dead stop to 60 mph in 12.3 seconds, quicker than the average gasoline-powered car. Yet they have an EPA city fuel economy rating of 27 mpg. Just thank Mercedes-Benz for the most powerful diesel passenger car engine you can buy: their 120 horsepower five cylinder turbo diesel.

The 300D is a four door sedan like the 300SD, but 11.7 inches shorter and 200 pounds lighter. The 300D has two cubic feet less interior room and three cubic feet less trunk space than the bigger SD. Though not space efficient in terms of overall length, the 300D offers a good combination of interior room, fuel economy, and acceleration.

The smaller, shorter two door 300CD is more difficult to justify. It has seven cubic feet less interior room, yet weighs the same as the four-door and returns no better EPA fuel economy. Its aerodynamic drag is only slightly improved over the four-door. In comparison with the Ford Escort Turbo GT, it is two feet longer, 1470 pounds heavier and has the same interior room, and 40% less trunk space, gets worse fuel economy, and is much slower and less maneuverable. The Mercedes costs four times as much as the Escort Turbo. Nonetheless, there will always be those who lust after the CD's sleek lines and teutonic solidity.

In Mercedes' defense, both the 300D and CD offer a substantial list as standard equipment, including air conditioning, central locking system, cruise control, power windows, electronic rear window defroster, first-aid kit, fog lights, four wheel disc brakes, AM/FM cassette stereo with automatic motorized antenna, and alloy wheels. And of course Mercedes-Benz quality and luxury are among the best.

Mercedes' smaller four-door turbo-diesel, the 300D, is the quickest diesel car you can buy.

Mercedes' two-door 300CD is a somewhat less efficient package than the 300D, but features uncommon style.

Model: Mercedes-Benz 300D

Importer: Mercedes-Benz of North America
One Mercedes Drive
Montvale, NJ 07645

Model year (s):   1982—(continuing)
Specifications for: 1983 Mercedes-Benz 300D
Vehicle type: front engine, rear drive
Body styles: 4-door sedan

Wheelbase: 110 in
Length: 190.9 in
Width: 70.3 in
Height: 56.6 in
Weight: 3580 lbs
SAE volume, interior/trunk: 92/12
Fuel tank capacity: 21.1 gal

Engine type: 5 in-line turbo diesel
Displacement: 2998cc/182.9 cu in
Turbo: AiResearch with integral wastegate
Maximum boost: 11 psi
Compression ratio: 21.5:1
Fuel system: Bosch diesel injection
Horsepower: 120 @ 4300 rpm

Transmission (s):   4-speed auto

Front suspension: ind, control arm, coil springs
Rear Suspension: ind, coil springs

Acceleration, 0–60 mph: 12.3 sec
Top speed: 107 mph
Braking, 70–0 mph: 208 ft
EPA Fuel economy, city/hwy: 27/33

Model: Mercedes-Benz 300CD

Importer: Mercedes-Benz of North America
One Mercedes Drive
Montvale, NJ 07645

Model years: 1982–(continuing)
Specifications for: 1983 Mercedes-Benz 300CD

Body styles: 2-door sedan

Wheelbase: 106.7 in
Length: 187.5 in
Width: 70.3 in
Height: 54.9 in
Weight: 2580 lbs
Fuel tank capacity: 21.1 gal

Engine type: 5 in-line turbo diesel
Displacement: 2998cc/182.9 cu in
Turbo: AiResearch with integral wastegate
Maximum boost: 11psi
Compression ratio: 21.5:1
Fuel system: Bosch diesel injection
Horsepower: 120 @ 4300 rpm

Transmission (s):   4-speed auto

Front suspension: ind, control arm, coil springs
Rear Suspension: ind, coil springs

Acceleration, 0–60 mph: 12.3 sec
Top speed: 109 mph
Braking, 70–0 mph: 208 ft
EPA Fuel economy, city/hwy: 27/33

MITSUBISHI TURBO DIESEL PICK-UP

For maximum fuel economy with average acceleration, the Mitsubishi Turbo Diesel is number one. It takes 14.6 seconds to get to 60 mph, but it returns 34 mpg in the EPA city cycle, versus the 2.6 liter's 23 mpg. In fact, the only pick-up truck with a better mileage rating than the Mitsubishi Turbo Diesel is Volkswagen's diesel—but the Volkswagen takes an agonizing 21 seconds to get to 60 mph, and has a 390 pound lighter payload capacity and a nine inch shorter bed than the Mitsubishi. With an EPA city rating of 30 mpg, the Mitsubishi four wheel drive Turbo Diesel Pick-up holds the four wheel drive fuel economy title. In fact, it gets even better mileage than Subaru's four wheel drive sedans and outaccelerates the normally aspirated Subaru to boot!

The 2.3 liter Mitsubishi turbo diesel engine puts out 80 horsepower. It uses a Mitsubihi TC05 turbocharger, which weighs only twelve pounds. A bosch VE fuel injection

Mitsubishi's turbo diesel pick-up offers an excellent combination of fuel economy and power.

The 2.4 liter Mitsubishi turbo diesel has a counter-rotating balance shaft to reduce vibration. Mitsubishi's turbocharger weighs a mere 12 pounds.

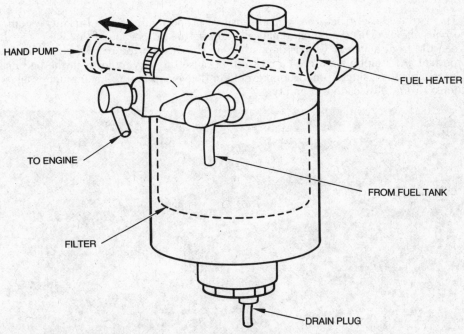

The Mitsubishi turbo truck has an all-in-one fuel filter/water separator/fuel heater. It's located at the top of the engine compartment for easy maintenance.

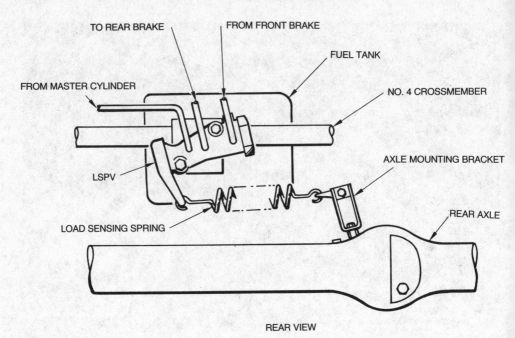

TO REAR BRAKE

FROM FRONT BRAKE

FUEL TANK

FROM MASTER CYLINDER

NO. 4 CROSSMEMBER

LSPV

AXLE MOUNTING BRACKET

LOAD SENSING SPRING

REAR AXLE

REAR VIEW

A unique load-sensitive proportioning valve on Mitsubishi turbo trucks adjusts rear brake line pressure depending on load to reduce skidding.

pump squirts the fuel. Two counter-rotating balance shafts reduce vibration. Mitsubishi claims the Quick Glow system heats the glow plugs in a mere 1.3 seconds at 32°F. An automatic fuel filter/water separator/heater keeps the diesel fuel temperature over 37°F.

The Turbo Diesel truck uses a load-sensitive proportioning valve braking system on the rear wheels. This system automatically adjusts the proportion of braking power between the front and rear wheels depending on the load on the rear wheels. This is an important and unique feature on a pick-up, because the varying loads in the bed make it impossible for a single proportioning ratio to deliver optimum braking under all conditions—from fully loaded to empty.

Dodge Ram 50 Turbo Diesel pick-up is virtually identical to Mitsubishi's, and Dodge dealers abound.

Dodge's version of the Mitsubishi 4WD turbo-diesel pick-up. You can't get better fuel mileage in a 4WD pick-up.

A standard 18 gallon fuel tank allows a cruising range of over 600 miles at the EPA city rating of 34 mpg. Tilt steering wheel and a cargo light are also standard. The four wheel drive version has part-time four wheel drive and locking/free-running front hubs for reduced drag and better fuel economy. The 4WD comes with all-terrain 15″ diameter tires, while the 2WD gets steel-belted radials.

The Mitsubishi turbo diesel truck is also sold by Dodge as the Ram 50 (2WD) and Power Ram 50 (4WD). Except for having different trim, the Dodge versions are identical to the Mitsubishis.

Model: Mitsubishi Turbo Diesel Pick-Up

Importer: Mitsubishi Motor Sales of America
10540 Tablert Street
Fountain Valley, CA 92708

Model year (s):   1984—(continuing)
Specifications for: 1984 Mitsubishi Turbo Diesel Pick-Up
Vehicle type: front engine, rear drive/4 wheel drive
Body styles: 2-door pick-up

Wheelbase: 109.4 in
Length: 184.4 in
Width: 65.0 in
Height: 59.8 in
Weight: 2640 lbs
Fuel tank capacity: 18 gal

Engine type: 4 in-line turbo diesel
Displacement: 2346cc/143.2 cu in
Turbo: Mitsubishi TCO5 with integral wastegate
Maximum boost: not available
Compression ratio: 21:1
Fuel system: Bosch VE diesel injection
Horsepower: 80 @ 4200 rpm

Transmission (s):   5-speed manual

Front suspension: ind., coil springs
Rear Suspension: rigid axle, leaf springs

Acceleration, 0–60 mph: 14.6 sec
Top speed: 80 mph
Roadholding: .66g
EPA Fuel economy, city/hwy: 34/45

PEUGEOT 505 TURBO DIESEL

Peugeot's best seller in the U.S., the 505 Turbo Diesel, is best known for its smooth ride, luxurious interior, and sturdy construction. It is the antithesis of a performance car, with poor acceleration, top speed, braking distances, and road holding. Yet the turbo diesel easily outaccelerates the "slow as a city bus" normally-aspirated 505 diesel, while also getting slightly better EPA fuel economy, an all-encompassing victory for turbocharging.

Peugeot's 2.3 liter diesel was designed from scratch as a diesel, not converted over from a gasoline design like many automotive diesels. The XD2S turbo motor has eighteen per cent more horsepower than the standard diesel. An AiResearch turbocharger pumps a maximum of 8.7 psi of air into the intake manifold before the wastegate opens. The compression ratio is down two points from the normally-aspirated diesel to 21:1. A Bosch VE mechanical injection pump squirts in diesel fuel. The pushrod-equipped block is cast iron. An aluminum cylinder head sports swirl combustion chamber inserts.

The Peugeot's unremarkable body was designed by the Italian stylist Pininfarina. It looks stodgy but not stately. It boasts excellent driver visibility, but a poor ratio of interior room to overall length. Suspension is fully independent, and a limited-slip differential improves rear wheel traction. Two common complaints about the Peugeot diesels are obtrusive engine noise and slow glow plug starting.

Celebrating their 25th year in the United States, Peugeot is offering a limited production Silver Edition for 1983. Fewer than 500 will be produced with the turbodiesel engine. Silver metallic paint, leather upholstery, heated driver's seat, power sunroof, Michelin TRX tires mounted on aluminum wheels, and Bosch fog lights are included in this package.

No performance car here, the Peugeot 505 Turbodiesel boasts a smooth ride, luxurious interior, and sturdy construction.

A victory for turbocharging: Peugeot's turbo diesel has 18% more power than their normally-aspirated version, and gets better fuel economy too.

Model: Peugeot 505 Turbodiesel

Importer: Peugeot Motors of America
One Peugeot Plaza
Lyndhurst, NJ 07071

Model year (s): 1981—(continuing)
Specifications for: 1983 Peugeot 505 Turbodiesel
Vehicle type: front engine, rear drive
Body styles: 4-door sedan

Wheelbase: 107.9 in
Length: 186.7 in
Width: 68.4 in
Height: 56.4 in
Weight: 3250 lbs
SAE volume, interior/trunk: 90/11
Fuel tank capacity: 18.0 gal

Engine type: 4 in-line turbo diesel
Displacement: 2304cc/140.6 cu in
Turbo: AiResearch TA0302 with integral wastegate
Maximum boost: 8.7 psi
Compression ratio: 21:1
Fuel system: Bosch VE mechanical diesel injection
Horsepower: 80 @ 4150 rpm

Transmission (s): 5-speed manual, 3-speed auto

Front suspension: ind, MacPherson strut, coil springs
Rear Suspension: ind, coil springs

Acceleration, 0–60 mph: 14.9 sec

Top speed: 94 mph
Roadholding: .69g
Braking, 70–0 mph: 214 ft
EPA Fuel economy, city/hwy: 28/36

PEUGEOT 604 TURBO DIESEL

The 604 is just a little more Peugeot than the 505. It is five and a half inches longer than the 505, and over an inch wider, yet the 604 has only one cubic foot more interior room than the 505. By comparison, the Dodge Aries and Plymouth Reliant are sixteen inches shorter than the 604, yet boast five cubic feet more interior room.

The 604 Turbodiesel and 505 Turbodiesel use the same engine and drivetrain. Acceleration and top speed of the heavier (250 lbs.) 604 is worse than that of the already slow 505. In fact, with a top speed of 88 mph and a 0–60 mph acceleration time of 17 seconds, the 604 is the slowest turbocharged car on the market, and one of the slowest cars available in the U.S. today. However, Peugeots enjoy a reputation for reliability and longterm durability that is backed up by Peugeot's thorough testing of every car off the production line.

The four wheel independent suspension with MacPherson struts up front and trailing arms with coil springs in the rear is similar to that of the 505. Michelin TRX tires are standard on the 604 though, improving braking and roadholding slightly over the base 505.

The 604 also offers as standard equipment air conditioning and a one step locking system: a single turn of the key locks all four doors, closes the windows and sunroof, and even locks the fuel tank. Options include an automatic transmission (5-speed manual is standard), metallic paint, and leather upholstery with a heated driver's seat.

Despite its prodigious length, the EPA considers the 604 a compact: not as much interior room as you might hope for in a sedan this big.

LUBRICATION SYSTEM

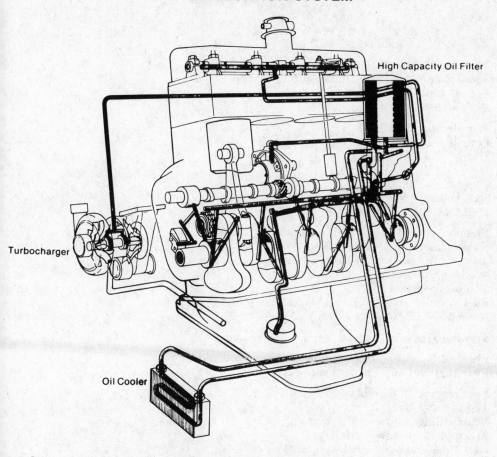

High Capacity Oil Filter

Turbocharger

Oil Cooler

PISTON INNER SKIRT COOLING

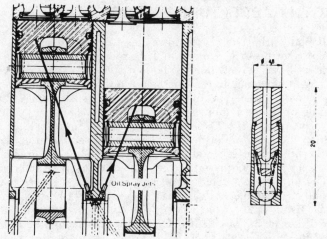

Oil Spray Jets

Peugeot's diesel was designed as a diesel from scratch, not converted from a gasoline engine. Piston inner skirt oil cooling, a high capacity oil filter and oil cooler add reliability to the turbo version.

Model: Peugeot 604 Turbodiesel

Importer: Peugeot Motors of America
One Peugeot Plaza
Lyndhurst, NJ 07071

Model year (s):   1981—(continuing)
Specifications for: 1983 Peugeot 604 Turbodiesel
Vehicle type: front engine, rear drive
Body styles: 4-door sedan

Wheelbase: 110.2 in
Length: 192.3 in
Width: 69.7 in
Height: 56.2 in
Weight: 3500 lbs
SAE volume, interior/trunk: 91/14
Fuel tank capacity: 18.0 gal

Engine type: 4 in-line turbo diesel
Displacement: 2304cc/140.6 cu in
Turbo: AiResearch TA0302 with integral wastegate
Maximum boost: 8.7 psi
Compression ratio: 21:1
Fuel system: Bosch VE mechanical diesel injection
Horsepower: 80 @ 4150 rpm

Transmission (s):   5-speed manual, 3-speed auto

Front suspension: ind, MacPherson strut, coil springs
Rear Suspension: ind, coil springs

Acceleration, 0–60 mph: 17.0 sec
Top speed: 88 mph
Roadholding: .71g
Braking, 70–0 mph: 210 ft
EPA Fuel economy, city/hwy: 28/36

VOLKSWAGEN RABBIT TURBO DIESEL

Volkswagen's Rabbit Turbo Diesel is both the least expensive and most fuel economical turbo diesel car on the U.S. market. Its EPA fuel economy matches the five-speed normally aspirated diesel Rabbit, though the four-speed normally aspirated diesel gets better mileage. With automatic transmissions, the turbo diesel gets better mileage than the normally aspirated diesel. The turbo diesel Rabbit gets 43 EPA city mpg, versus 32 mpg for the best gasoline Rabbit.

Acceleration-wise, the Turbo Diesel accelerates to 60 mph in 15.0 seconds, 3 seconds quicker than the basic diesel, but 3.5 seconds slower than the fuel injected gasoline version. So there's a place for all three Rabbit engines (not to mention the quick GTI), depending on your fuel preferences and acceleration vs fuel economy vs initial cost outlook. The normally-aspirated diesel has only 52 horsepower, while the Turbo Diesel has 68 and the fuel/injected gasoline engines have 74 horsepower.

Of course, the Rabbit is the front wheel drive compact that started the front wheel drive econobox craze of the late 1970s. Originally manufactured in Germany, it has been assembled in Pennsylvania since 1978. While there are plenty of good and less expensive econoboxes from Japan and the U.S., the Rabbit is still one of the best. It has excellent brakes and handles well, although it's not up to the latest standards aerodynamically. The Rabbit is smaller than it looks. Many people think of it as a Dodge

Optimum fuel economy in a turbo-diesel? No other choice than VW's turbo diesel Rabbit, discontinued at the end of '83.

Omni/Ford Escort-sized car. It is actually about ten inches shorter than those two cars, and has six or seven cubic feet less interior room.

Despite its impressive performance claims, the Rabbit Turbo Diesel has not been a big seller, and most Rabbit buyers who want a diesel have opted for the less expensive, higher mileage normally aspirated diesel. As a result, Volkswagen cancelled the Rabbit Turbo Diesel at the end of 1983.

Model: Volkswagen Rabbit Turbo Diesel

Manufacturer: Volkswagen of America
27621 Parkview Blvd.
Warren, MI 48092

Model year (s):   1983
Specifications for: 1983 Volkswagen Rabbit Turbo Diesel
Vehicle type: front engine, front drive
Body styles: 3-door sedan, 5-door sedan

Wheelbase: 94.5 in
Length: 153.3 in
Width: 63.4 in
Height: 55.5 in
Weight: 1970 lbs
SAE volume, interior/trunk: 78/14
Fuel tank capacity: 11 gal

Engine type: 4 in-line turbo diesel
Displacement: 1588cc/196.9 cu in
Turbo: AiResearch with integral wastegate
Maximum boost: 10.2 psi
Compression ratio: 23:1
Fuel system: diesel injection
Horsepower: 68 @ 4500 rpm

Transmission (s):   5-speed manual, 3-speed auto

Front suspension: ind, MacPherson strut, coil springs
Rear Suspension: ind, coil springs

Acceleration, 0–60 mph: 15.0 sec
Top speed: 95 mpg
Roadholding: .73g
Braking, 70–0 mph: 195 ft
EPA Fuel economy, city: 43

VOLKSWAGEN JETTA GL TURBO DIESEL

The Jetta sedan's layout is less space—efficient than the Rabbit's hatchback design. They share the same height and width, but the Rabbit is fourteen inches shorter. Interior dimensions are the same on the 2-door, and the Rabbit has slightly more interior room if you're comparing 4-doors. The Rabbit has slightly more trunk space (yes, that's with the rear seat up). But if you want or need that separate trunk, the Jetta is a high fuel economy, low buck alternative to an Audi 4000 or BMW 318.

Superb fuel economy and average acceleration can be had in the Jetta GL with the optional Turbo Diesel engine. The Turbo Diesel gets 41 EPA city mpg, 2 miles per gallon less than with the normally-aspirated diesel. EPA highway mileage is actually 1 mpg better with the turbo. Top speed is up to 94 mph, 7 mph higher than the base diesel. 0–60 mph acceleration is cut by 3 seconds with the turbo. While acceleration is nowhere near gasoline-powered Jetta territory, the Turbo Diesel gets 11 more EPA city mpg than the gasoline version. But the Turbo Diesel Jetta GL costs $1000 more than the gasoline-powered Jetta GL.

The Turbo Diesel's cylinder block has more cast iron than the normally-aspirated diesel. The piston wrist pins are hardened, and valve seats, piston rings and exhaust valves are all upgraded to withstand the heat of turbo duty. The cylinder head itself is made of higher sodium-content aluminum. A spray nozzle squirts oil at the underside of each piston for cooling. The turbo motor only weighs sixteen pounds more than the normally-aspirated diesel. The Turbo Diesel Jetta is available with an automatic or five-speed manual transmission. For '84 the Turbo Diesel is not available in the 2-door Jetta.

Like the Rabbit but need a separate trunk? VW's Jetta fills the bill, but it's not as light or space-efficient as the hatchback Rabbit.

For '84, there's no Rabbit turbo diesel, so the way to go is with a Jetta turbo diesel. The turbo gets 1 mpg better mileage on the EPA highway cycle than the normally-aspirated diesel.

Model: Volkswagen Jetta GL Turbo Diesel

Importer: Volkswagen of America
27621 Parkview Blvd.
Warren, MI 48092

Model year (s):   1983—(continuing)
Specifications for: 1984 Volkswagen Jetta GL Turbo Diesel
Vehicle type: front engine, front drive
Body styles: 4-door sedan

Wheelbase: 94.5 in
Length: 167.8 in
Width: 63.4
Height: 55.5 in
Weight: 2190 lbs
SAE volume, interior/trunk: 77/13
Fuel tank capacity: 10.6 gal

Engine type: 4 in-line turbo diesel
Displacement: 1588cc/196.9 cu in
Turbo: AiResearch with integral wastegate
Maximum boost: 10.2 psi
Compression ratio: 23:1
Fuel system: diesel injection
Horsepower: 68 @ 4500 rpm

Transmission (s):   5-speed manual, 3-speed auto

Front suspension: ind, MacPherson strut, coil springs
Rear Suspension: ind, coil springs

Acceleration, 0–60 mph: 15.5 sec
Top speed: 94 mph
Roadholding: .73g
Braking, 70–0 mph: 210 ft
EPA Fuel economy, city/hwy: 41/54

VOLKSWAGEN GL TURBO DIESEL QUANTUM/QUANTUM WAGON

The Quantum Turbo Diesel sedan gets the best EPA fuel mileage of any sedan of its interior size—by a healthy margin. Its closest competition is Renault's Alliance, which has roughly the same interior size as the Quantum, and gets 37 EPA city mpg on gasoline, as opposed to the Quantum's 41 mpg on diesel. Despite the turbo, the Quantum accelerates much slower than the Alliance, though the Quantum has a 3 mph advantage in top speed. But you could buy two Alliances for the price of one Quantum Turbo Diesel. The Quantum has one performance strength few people are aware of, however: It's one of the best braking cars you can buy. Only a few sports cars, like Porsche's 911 and 928 and the '84 Corvette, can stop in a shorter distance.

There's only one station wagon you can buy that gets better mileage than Volkswagen's Quantum Turbo Diesel Wagon, and that's Nissan's miniscule Sentra diesel wagon, which beats the Quantum by a mere 1 mpg on the EPA city cycle. If you need more room for your wagoning than the Sentra can provide, and high fuel mileage is your game, "Quantum Turbo Diesel Wagon" is the name. There's no other station wagon in its class.

For '84, both the Quantum Sedan and Wagon come with power windows, central locking system, electrically adjustable heated outside mirrors, air conditioning, cruise

The Quantum Turbo Diesel offers by far the best fuel mileage of any sedan of its interior size. It has super brakes and an impressive list of standard features.

When you look at fuel mileage and interior room in a station wagon, the Quantum Turbo Diesel Wagon is in a class by itself.

VW's turbo diesel weighs only 16 pounds more than their normally-aspirated diesel, but puts out 30% more power.

control, alloy wheels, and a four speaker AM/FM auto-reverse cassette stereo system (with power antenna) as standard equipment. That's about $2500 worth of options on most cars. Also standard on the Wagon is an aerodynamically-designed roof rack (no more whistle) and a rear window wiper. The Wagon has a 60/40 split rear seat and a luggage cover for the cargo area—good for security from prying eyes.

Model: Volkswagen GL Quantum

Importer: Volkswagen of America
27621 Parkview Blvd.
Warren, MI 48092

Model year (s): 1983—(continuing)
Specifications for: 1984 Volkswagen GL Quantum
Vehicle type: front engine, front drive
Body styles: 4-door sedan

Wheelbase: 100.4 in
Length: 180.2 in
Width: 66.9 in
Height: 55.1 in
Weight: 2579 lbs
SAE volume, interior/trunk: 89/12
Fuel tank capacity: 15.8 gal

Engine type: 4 in-line turbo diesel
Displacement: 1588cc/196.9 cu in
Turbo: AiResearch with integral wastegate
Maximum boost: 10.2 psi
Compression ratio: 23:1
Fuel system: diesel injection
Horsepower: 68 @ 4500 rpm

Transmission (s):   5-speed manual, 3-speed auto

Front suspension: ind, MacPherson strut, coil springs
Rear Suspension: ind, coil springs

Acceleration, 0–60 mph: 17.0 sec
Top speed: 92 mph
Braking, 70–0 mph: 181 ft
EPA Fuel economy, city/hwy: 41/50

Model: Volkswagen GL Quantum Wagon

Manufacturer: Volkswagen of America
27621 Parkview Blvd.
Warren, MI 48092

Model year (s):   1983—(continuing)
Specifications for: 1984 Volkswagen GL Quantum Wagon
Vehicle type: front engine, front drive
Body styles: 4-door wagon

Wheelbase: 100.4 in
Length: 183.1 in
Width: 66.9 in
Height: 55.1 in
Weight: 2600 lbs
SAE volume, interior/trunk: 89/38
Fuel tank capacity: 15.8 gal

Engine type: 4 in-line turbo diesel
Displacement: 1588cc/196.9 cu in
Turbo: AiResearch with integral wastegate
Maximum boost: 10.2 psi
Compression ratio: 23:1
Fuel system: diesel injection
Horsepower: 68 @ 4500 rpm

Transmission (s):   5-speed manual, 3-speed auto

Front suspension: ind, MacPherson strut, coil springs
Rear Suspension: ind, coil springs

Acceleration, 0–60 mph: 17.1 sec
Top speed: 92 mph
Braking, 70–0 mph: 181 ft
EPA Fuel economy, city/hwy: 41/50

VOLVO 760 GLE TURBO DIESEL

Volvo's 760 Turbo Diesel is an alternative to the Mercedes 300D Turbo Diesel—an alternative that costs $12,000 less. The Volvo's acceleration is tied with the Mercedes—they are the quickest diesel cars you can buy. While the Mercedes has a four mile per hour edge in top speed, the Volvo gets one mile per gallon better EPA city fuel economy, and produces seven more mpg on the highway. The Volvo has slightly more interior room, and a whopping five cubic feet more trunk space. Yet the Volvo is two and a half inches shorter and one and a half inches narrower than the Mercedes.

Though the 760 appears to be larger than the standard 240 series Volvos, it is actually a half inch shorter and seven-tenths of an inch lower than the 240 sedans. Yet the 760 boasts a five inch longer wheelbase, which reduces pitching and aids high speed

Vovlo's 760 Turbo Diesel compares very favorably with Mercedes' 300D—at a considerably lower price.

The 760's 2.4 liter six-cylinder turbo diesel combines 28 EPA city mpg fuel economy with 12.3 second 0–60 mpg acceleration.

stability. Most importantly, the 760 has four cubic feet more interior room and two cubic feet more trunk space than the 240. While the 760 is a big improvement over the 240, its space efficiency is poor in comparison with domestic front wheel drive sedans.

The 760's 103 horsepower engine is made by Volkswagen. It is an in-line six cylinder with a cast iron block and aluminum head. As with all automotive diesels, the diesel fuel begins to burn in a small precombustion chamber with a relatively rich mixture before spreading through a swirl passage to the main combustion chamber, where a leaner mixture reduces exhaust emissions. The Garrett T03 turbocharger's integral wastegate allows up to ten pounds of boost. The mandatory four-speed manual transmission has an electrically-operated overdrive.

The 760 offers a good combination of luxury, comfort, acceleration, and fuel economy. Its handling is mediocre. Though a bargain in comparison with a Mercedes, the Volvo's cost is high in comparison with domestic sedans such as the Dodge 400/600 and Pontiac 6000. The Volvo's base price does include such extras as air conditioning, cruise control, electric windows, and a driver's seat heater, though.

Model: Volvo 760 GLE Turbo Diesel

Importer: Volvo of America Corp.
                 Rockleigh, NJ

Model year (s):   1983—(continuing)
Specifications for: 1983 Volvo 760 GLE Turbo Diesel
Vehicle type: front engine, rear drive
Body styles: 4-door sedan

Wheelbase: 109.1 in
Length: 188.4 in
Width: 68.9 in
Height: 55.5 in
Weight: 3155 lbs
SAE volume, interior/trunk: 93/17
Fuel tank capacity: 21.6 gal

Engine type: 6 in-line turbo diesel
Displacement: 2383cc/145.4 cu in
Turbo: Garrett T03 with integral wastegate
Maximum boost: 10 psi
Compression ratio: 23.0:1
Fuel system: Bosch VE diesel injection
Horsepower: 103 @ 4800 rpm

Transmission (s):   4-speed manual with electrically-operated overdrive

Front suspension: ind, MacPherson strut, coil springs
Rear Suspension: rigid axle, coil springs
Acceleration, 0–60 mph: 12.3 sec
Top speed: 103 mph
Roadholding: .73g
Braking, 70–0 mph: 206 ft
EPA Fuel economy, city/hwy: 28/40

5
Who's Who in Turbo Kits

Purchasing a turbo kit can be an exciting experience that is often accompanied by dreams and visions of transforming your anemic "set-of-wheels" into a powerhouse that could challenge Big Daddy's Don Garlits' jet-engined drag racer. On the other hand, purchasing a turbo system kit can also prove frustrating and unnecessarily more costly than it should be, if the kit purchaser is not in tune with this rapidly changing aftermarket performance industry. To help you avoid frustration and unnecessary cost, this chapter reviews the major turbo manufacturers and their kits. In order to avoid redundant comments on standard features offered by virtually all kit manufacturers and to encourage you to ask questions about features which may not be addressed in the literature supplied by a particular kit manufacturer, the following points should be considered when reading the individual turbo kit reviews. They are:

• Some kits are not designed for use with air conditioning, power steering, or other accessories.
• Emission devices may have to be altered in some cases.
• EPA/CARB (California Air Resources Board) certification may not exist and unless stated otherwise, kits are not legal for street use in California.
• Some systems may not be suitable for automatic transmission applications.
• Some systems may require drivetrain modifications.
• Most systems are hardware complete, however, exhaust system modification and welding may be necessary.
• Most systems do not require body or chassis modifications.
• Boost gauges may not be standard equipment.
• Many systems do not have express warranties.
• Is intercooling offered? This feature of some turbo kits reduces both heat and pressure stress in the engine, substantially increases system performance (sometimes, in part, by permitting more boost pressure), and helps substantially to prevent detonation.

If there's one point this chapter's introduction should get across, it's the importance of selecting a turbo kit on the basis of quality and not just price. A poor quality kit can require months of time and thousands of dollars of owner/installer development to work properly and reliably. A poor quality kit can blow up your motor, costing the owner/installer at least a thousand dollars, not to mention the down time. A quality kit, on the other hand, may cost five hundred or a thousand dollars more, but the far superior performance and reliability, not to mention reduced installation and development costs, will make the quality kit the real bargain everytime.

A wastegate is of critical importance to protect the engine from an overboost condition and to allow power-producing boost at low engine speeds. The few kits still on the market which don't have wastegates should be avoided. Quality turbo kits use custom exhaust manifolds, not stock manifolds with welded-on adapters. Custom cast intake manifolds are usually not critically important, but they are a plus, generally providing improved performance and driveability. They also offer reduced intake temperatures for reduced engine stress and detonation.

Quality kits also offer boost-activated ignition retard systems. Water injection can be a plus or a minus, depending on your view of their performance and hassles. Survey the fuel system, including fuel enrichment systems, auxiliary fuel pump, carburetor and/or jetting changes. Spare jets, needed for fine tuning the carburetor, are cheap, but finding a source for them can be a super headache.

Some kits are made specifically for one model of car, while others are designed for a number of vehicles using the same basic engine. These multi-vehicle kits usually require owner/installer adaptations for individual models and years. These adaptations can range from minor filing and drilling to manufacturing mounting brackets to replacing the entire ignition system. Look before you leap. Some turbo kits designed for specific models even include the exhaust system.

Intercoolers are expensive, but they are a big plus in any turbo installation, improving horsepower while reducing the thermal strain on the engine. Some kits include low compression pistons, which are a must if you have a high compression engine. Most kits, but not all, include boost gauges. Strangely, few kits include oil coolers, though these are standard fare on most factory turbocars. Some kits are more hardware—complete than others—this may sound minor, but the cost of hose clamps, metric bolts, and fuel lines and braided oil lines can add up, as can the time spent shuttling to hardware and auto parts stores.

Any engine destined for a turbo kit must be in prime condition. Excellent compression test results are mandatory, and any oil burning problems must be fully resolved. After all, most factory turbo cars use special valves, pistons, gaskets, rods, cranks, bearings, fasteners, and cooling and oiling systems to handle turbo duty. If your normally-aspirated engine isn't in prime condition, it doesn't stand a chance under even mild boost. A thorough engine rebuild is often the best first step for turbokitting, and the low compression pistons, high output oil pump, and heavy duty rod bolts sometimes substituted for stock parts can be engine life savers later.

One intangible ingredient of turbokit installation that cannot be overstressed is having a good attitude. Do the job like it was being recorded on film for posterity. Think ahead. Have all the necessary parts, lubricants, parts bins and tools ready. Work in a clean, dry, temperate garage with a well-swept floor and adequate light. This way you'll be less likely to make mistakes or lose pieces.

Wear work clothes, and don't work under the pressure of time, if you can help it. All this will minimize difficulties, help you keep your cool, and get the job done in a manner that will head off potential problems. A turbo-kit installation is no small project, and there are a myriad of things that can go wrong even if you're careful. So do the job right the first time, and resist taking that first test drive until all the details have been finished.

One thing to remember after you do your conversion is that turbocharged engines are harder on oil than normally aspirated ones. The primary culprit is heat—especially high temperatures encountered right after engine shutdown in the turbo's bearing. You can counteract this effect somewhat by idling your engine for a few minutes after a hard run, before you actually turn it off. But oils with extra detergents and anti-oxidants, and somewhat less viscosity index improvers will withstand the heat and corrosion in the turbo much better.

You can decrease the amount of "VI" improvers in the oil by opting for 15W-40 or 10W-30 oil instead of 10W-40. Valvoline makes "Turbo V", an oil designed expecially for turbo service, and other manufacturers are sure to follow suit. Generally speaking, better quality oils will have extra antioxidants and detergents. Changing oil and filter more frequently will help, too. Enjoy!

AK MILLER

The Ak Miller people waste no time in offering the potential turbo system customer "Some Important Facts to Consider!" Facts that make no turbo bones about the "favorable power-per-dollar ratio" a turbo system has over "conventional hop up methods which would provide a similar power boost" in normally aspirated engines. In fact, the Ak Miller people approximate the costs of turbocharging at 1/3 that of going the n/a route for similar power gains. Speaking of power gains, Ak Miller's highly engineered kits, sporting AiResearch turbos, have competition credits ranging from a 300 mph sports car to a variety of drag strip records. Ak Miller's competition exploits may not yet have peaked, but one venerable peak in competition has been reached: Colorado's Pike's Peak. Several assaults have been launched on this peak with propane and an Ak Miller turbo-propane system extracting every BTU available for conversion into horsepower.

Propane conversion kit development and industrial powerplant turbocharging experience definitely makes Ak Miller unique in the turbo system kit manufacturing business. However, even with his great diversity of turbocharging experience the product line-up favors Fords with six kits offering everything from a blow-through 302 C.I.D. kit to a 2000 cc Pinto/Capri kit. Also covered are 2600 and 2800 Capris, 360/390 trucks, 351M/400, Courier 2300, 170/250 sixes, 240/300 sixes, and the 351 Pantera. Chevrolet is also graced with Ak's well engineered power-on-demand, power elixirs including the twin turbo big and small block applications, the 292 six, and Ak's most popular kit, the small block Chevrolet draw-through system. Ak Miller turbocharger conversions are designed to use stock carburetion in most cases. However, custom/special application carburetor adapters (Holley included) are offered along with universal system compo-

Ak Miller blow-through turbo-kitted late model 302 Mustang sports an Akwood intercooler. The now-functional hood scoop directs outside air through intercooler.

Ak Miller Ford 302 blow-through kit uses an AiResearch T04 turbocharger.

Ak Miller's kit for the 2-liter Pinto/Capri uses an AiResearch T04 turbocharger in a draw-through installation.

Chevy small block blow-through kit by Ak Miller.

Ak Miller's in-line six-cylinder Ford kit.

This Ak Miller twin-turbo kit mounts easily to small block Chevies using a Holley bolt-pattern four barrel carburetor.

nents to accommodate the custom designer. Mopar small block V-8 turbo parts are also available. Boost control is accomplished in most kits by in-line pressure regulators (TC-2's) as opposed to an exhaust by-pass or wastegate valve system, but an optional Ak Miller Boost Guard boost controller is offered with the necessary adapter and wastegate valve.

Fuel injection technology and imported engines have not gone unnoticed at AME but the standard product line is virtually limited to one kit fitting in this category—one made for a Datsun 280Z.

Kit prices run from $968.00 to $1875.00 and generally contain all the basic components, though additional hardware, fittings, and oil lines are often necessary. Exhaust pipes and low restriction mufflers are required but must be purchased at additional cost. Throttle linkage, speed control, A/T kickdown and certain readily obtainable hardware pieces may also have to be purchased separately. The warranty bottom-line is "defects in materials and workmanship only" with freight costs paid by the customer if replacement is in order. Custom designed systems are available as well as individually purchased component parts to build your own system.

ARKAY

A 150 horsepower Holley-carbureted turbo system propelling a Mazda RX-7 from 0–60 mph in 6.9 seconds garners undeniable respect. Arkay and its owner Kas Kastner leave little doubt that they are into turbocharging "sporty little cars" in a big way when they design turbo systems to perform as this Mazda RX-7 does. This is no surprise, however, if one realizes that the distinguished Mr. Kastner spent years terrorizing sports car tracks and a few competitors with his aggressive driving style and well prepared sports cars—cars he prepared so well, in fact, that he later became U.S. Competition Director for Triumph automobiles.

RX-7 kit by Arkay uses a Holley four-barrel carburetor. Complete exhaust system is available.

Want more power for your TR-6? Call Kas Kastner at Arkay, the Triumph turbo expert.

Arkay's VW Dasher diesel kit shows typical Arkay thorough-
ness—from intake manifold to muffler.

Mazda 2200 truck kit by Arkay. This top quality kit even works with factory air conditioning.

This performance preparation experience has been carried forth into the turbo age with predictable results: well designed high performance turbo kits featuring Ai-Research, Rajay, or IHI turbochargers, wastegate boost control, cast aluminum or 16 gage steel induction tubing, water injection, and free-flow exhaust systems. An additional feature of the Arkay systems is that they are designed to "go-back-to-stock" very easily in the event that you wish to sell the car or meet particular inspection system requirements.

Diesel engine systems are available for the Rabbit/Dasher and Mazda pickup while fuel injected systems are available for the gasoline Scirrocco, Jetta, Rabbit, and the new GTI. The GTI kit includes a boost gauge, ten row aluminum oil cooler, and a boost-activated fuel enrichment system. Holley carburetion is used on the RX-7, a Weber 40 DCOE on the Triumph TR-7, and SUs on most other carbureted Arkay kits. In all, there are approximately 15 kits offered at present including three British cars (TR-6, TR-7, MGB), and four German cars (Scirrocco, Jetta, Rabbit, Dasher diesel). Mazda 626, Honda Prelude, Accord, and Civic, Subaru 1600 and 1800, Toyota Celica 20R, 22R and Toyota 4WD pickup are all covered. Diesel kits include the VW Rabbit/Dasher and Mazda diesel pickup. These kits range in price from $1195.00 to around $1800.00; however, certain often—suggested ancillaries such as water injection ($140.00) and exhaust systems ($150 plus) have to be tacked onto the basic kit price in some cases. Boost-activated water injection comes with the Subaru kit.

One additional device is offered, and might well be described as an insurance package. It is in the form of an eletronic detonation control unit dubbed the ADC ($325.00) by Arkay and should be given serious consideration when purchasing any turbo system. This unit controls detonation (the demonic poltergeist of turbocharged engines) by signalling the water injection pump to start pumping at a certain boost pressure and

Gasoline VW Rabbit kit by Arkay.

Not even sure it was a TR-8 that blew you off? Arkay has a turbo kit for your TR-7. With a little work, it might even fit a Saab 99.

Arkay's kit for the '83 Celica turns this super hander into a Supra-eater.

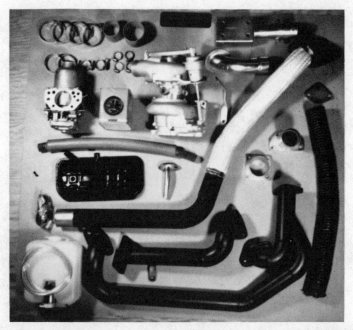

Arkay's Subaru kit even includes water injection. Note the cross-over exhaust header necessary for the unique flat four.

Arkay Subaru kit. The spare tire must be moved to the trunk.

Arkay's Honda Prelude kit is good for 8.8 second 0–60 mph acceleration.

Arkay's kit for the carbureted '82 Toyota Celica uses a two-barrel Solex carburetor for 7.4 second 0–60 mph acceleration.

then both increasing pump speed and retarding ignition timing linearly with increases in boost pressure. ADC is included in the Celica and VW gasoline kits. Arkay states in their literature that "performance is relative" and quite naturally the performance with Arkay systems and other brands tends to vary somewhat from system to system. Using 5-7 psi boost and 91 octane fuel, Arkay's Honda Prelude system, for example, showed 1–3 mpg improvements and chalked up 0–60 mph times of 8.8 seconds. Installation times will vary from system to system but most Arkay systems are designed as complete bolt-on systems and the company says the TR-6 system, for example, should take only 8–10 hours to install.

BAE

Car & Driver has proclaimed that BAE is "the world's largest emporium of turbocharging," a pretty hefty statement but one that is further supported by BAE's claims that they produce more turbocharger systems than all of their competition combined. Largest or not, BAE is well known and has been bathed in continued media attention at home and abroad to the tune of a hundred or more magazine articles on their systems and products. This is partly due to the fact that they offer more than 60 different systems for specific vehicles ranging from BMWs and DeLoreans to GMC and Toyota diesel pickups.

Two BMW 320i kits are available, one for '79 and earlier, another for '80 and later. This kit is CARB approved and incorporates sports chromed components. 528i, 530i, and 630i kits are available. BAE has three kits for the 633i/733i, two of which pass CARB standards and are offered with optional intercoolers.

Four Porsche 911 kits are available, as is a 924 kit. Four VW Rabbit/Jetta/Dasher kits are offered, including a CARB-approved version. Ferrari's 308 is covered, as are both single and twin cam Toyota Supras. An intercooler is optional on the twin cam. Kits are available for Datsun's 280Z and ZX. The Mazda RX-7 kit can be combined with BAE's HP System which includes a 4-barrel carb and air cleaner.

The BMW 320i is one nice sedan—but BAE's turbo kit will make it a quick sedan, too. Thanks to BMW's thoughtful separate cylinder-casting design, you don't have to pull the motor to install low compression pistons.

The Porsche 930 is but a memory in the U.S., but 911SC owners can get similar performance with this BAE kit. And a/c is a o.k.

BAE's Porsche 924 turbo kit.

The Rabbit Turbo Diesel is discontinued for '84, but BAE has early and late model Rabbit Diesels covered with this turbo kit.

On the domestic front, BAE has five different kits for small block Chevy pick ups and 4×4s. A Dodge/Chrysler 440 kit is for class A chassis motorhomes, and has CARB approval. Another 440 kit fits pickups and 4×4s. Do you have an AMC 258 six cylinder? No problem. A Delorean? No trouble here either. A Cadillac 4.1 liter? BAE factory installation only.

For diesels, BAE has kits covering: the International Scout and new 6.9L (Ford) diesels; Mercedes 220D, 240D, and 300D; VW Rabbit and Dasher diesels; Chevy and GMC 350 diesel pickup; Dodge/Mitsubishi diesel pickup; Peugeot 504D; 350 diesel powered Eldorado, Toronado, and Seville; Isuzu-powered Airstream motorhome, Isuzu and Chevette diesel, Isuzu and Chevy LUV pickup; Toyota 2.2 liter diesel pickups; Chevy's 6.2 liter diesel trucks; and VW's Vanagon diesel. What a list! These systems are mostly wastegate-controlled for maximum performance over a wide range of rpms. However, a non-wastegated system is offered for the fuel injected Datsun 280ZX, Mercedes Benz 300D, and several others.

Cosmetically, most BAE systems feature polished, cast aluminum intake manifolds and carburetor adaptors in conjunction with functional/cosmetic amenities such as aircraft-type oil lines and fittings and silicone intake hoses. Many systems will meet EPA and or CARB (California Air Resources Board) standards which may be of prime consideration to some buyers faced with periodic emission inspections. BAE maintains a sophisticated test lab in order to meet these standards and help guarantee top performance.

BAE's GM 6.2 liter diesel kit fits snugly in a Pace Arrow motor home. What a way to see the U.S.A.!

BAE kit for the '77–'81 Mercedes-Benz 300D—air conditioning and all.

While most systems are bolt-on units requiring no previous turbo experience and no special tools, BAE does offer component parts for customer designed installations. The kits start from $1400 with performance claims in the range of 20–70% HP increases for diesel engines and 40–100% for gasoline engines.

BANKS

If you wish to be afflicted by "a clear-cut case of excessive awesomeness" in the form of a 350 cubic inch Chevy engine developing an atomic reactor-like 400 plus horsepower, then a Gale Banks Engineering (GBE) turbo system might be your cup of fuel. GBE is undoubtedly a salient example of what specialization is all about—Chevrolet engine specialization to be specific. While the Banks people might be best known for their marine turbo miracles, they are just as adept out of the briny, if one considers a twin turbo (Rajay) blow-through system neatly perched topside of a Camaro/Firebird 350 cubic inch engine and developing 400 horsepower, as being adept. This performance can be achieved on pump gas and the whole package will maintain stock hood clearance to boot!

GBE turbo systems are wastegate-controlled (usually in the 8-12 psi range), cosmetically gorgeous with "custom rod" glitter and undeniably possessed of an aura of engineering and fabrication excellence second to none. State-of-the-art tricks also in evidence include the boost pressure-driven water injection, intercooling, water heated induction plumbing (for better low speed operation), and reverse deflection turbo pistons. An air box dump valve is designed to maintain turbo turbine speed by dumping

Gale Banks' complete and famous twin turbo small block Chevy kit features accessory drives and an optional a/c package.

Installed in a Corvette, the Banks' twin turbo kit has plenty of hood clearance. Better for aerodynamics than a GMC supercharger. Remote air filters mount in front of the radiator.

excessive boost pressure when the throttle is suddenly closed. This bleeding reduces compressor back pressure, allowing the turbine to spin more freely during shift periods, which translates into a more responsive system when boost is again summoned. It should be noted that GBE systems employ blow-through turbocharging which has to its credit such features as quick throttle response (read reduced turbo-lag), stock carburetor location and improved turbocharger efficiency by virtue of pumping only air into the carburetor, (as opposed to drawing the fuel mass through the turbo as with the draw-through turbo system which requires additional energy from the turbo).

GBE's dedication to engineering excellence, attention to details and cosmetic appointments do not come cheaply! Their famous Chevy small-block twin turbo kit, for example, scales in at around $3700.00, but one must also realize that these kits are truly complete for easy installation by the layman. Holley carburetion modified by Banks is the most often-used fuel metering device and in the case of the Chevy small-block kit it is perched atop the venerable Edelbrock Torker manifold (supplied if you don't already own one). Exhaust manifolding is definitely built to survive Chevy small-block punishment with heavy wall pipe supported by ½" thick flanges.

These kits are race-bred because GBE's heritage, as with many highly successful turbo system manufacturers, is deeply routed in racing, with their marine engines capturing national titles in Fuel Hydro, Grand National, Inboard Tunnel classes, and a five year dominance (5 World Championships) in the World Endurance River Racing Class.

Located in San Gabriel, California, the Gale Banks Engineering facility occupies in excess of 20,000 square feet and employs thirty people. Included in this complex is a complete engine machine shop, fabrication shop, production machine shop, engine and chassis dynamometers, engineering laboratory, manufacturing and administrative offices. Gale Banks has been contracted to consult, or provide product to, many major automotive and marine companies, including Buick Motor Division, Chevrolet Motor Division, Pontiac Motor Division, Lawless Detroit Diesel, Mercruiser Marine, Volvo America, Volvo Sweden, National Highway Traffic Safety Administration, and others.

All of Banks' complete engines are thoroughly dyno tested at their shop before being

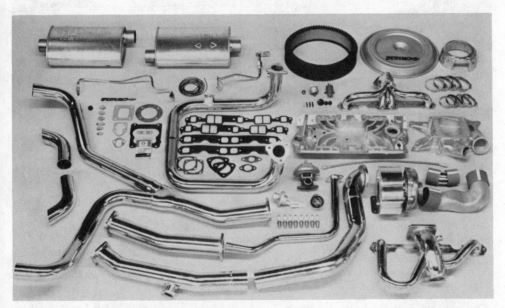

Gale Banks also sells single turbo kits for small block Chevy Corvettes and Camaros. These kits even include water injection and a complete exhaust system. Note the custom intake manifold and tubular steel headers. All the tubular steel parts are chromed—even the exhaust pipes!

delivered to the customer. This includes initial camshaft break-in and ring seating, certification of performance under load, and post-assembly dyno check of cylinder leakage. This testing ensures that each engine meets their specifications and will perform properly upon installation.

GBE will soon add a kit for the GM 6.2 liter diesel engine to its line-up, with everything provided from air cleaner to exhaust system. In addition to this new kit and the other fine kits, GBE also manufactures and sells separately turbo cams, turbo pistons, and other trick Chevrolet engine goodies.

BOB SHARP RACING

Bob Sharp Racing (BSR) has been designing, building, racing, and winning with turbocharged Datsun 280ZXs for years. It is no wonder that their street turbo systems possess the state-of-the-art turbo technology required to launch and maintain their highly successful racing efforts. The BSR Stage II turbo system, for example, features forged turbo pistons, 8.2:1 compression ratio O-ringed blocks, a balanced and blueprinted boost control, Mecca turbo oiler, and wastegate-controlled IHI turbochargers. Dyno tested BSR Stage II turbocharged Datsun 280 Z engines show their race heritage by pumping out 240 or more horsepower with only 10 psi of boost. The turbo system plumbing is factory-like in appearance and construction, and features heavy gauge steel tubing. Depending on individual customer driving styles and/or wishes, the BSR Stage II turbo systems can be tailored to meet the application, including the addition of special microprocessors and injectors for higher levels of fuel enrichment, special gauging and special turbo exhaust systems. Stage II turbo systems are sold as complete turbo-engine packages ready for installation by the customer or the people at Bob Sharp Racing. A trade-in (core) fee is offered and is on a sliding scale depending on the model

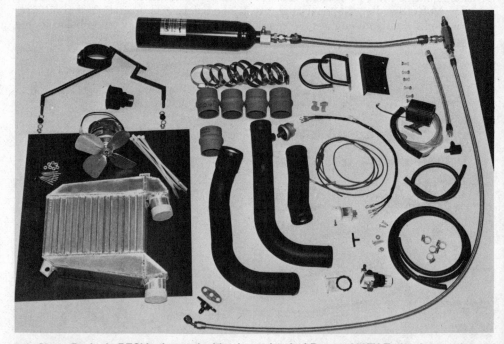

Bob Sharp Racing's BECkit almost doubles boost level of Datsun 280ZX Turbo. Intercooler and fan hide under left front fender. Mecca turbo oiler is included. Bob Sharp also offers complete turbo kits for 240, 260, and 280 Datsun ZS.

and condition of the Z or ZX engine to be traded. BSR Stage II turbo engine systems are currently available for the 240, 260, and 280 (Z&ZX) models plus custom-designed systems for the 4-cylinder Datsun/Nissan product line.

In addition to the Stage II turbo systems, Bob Sharp Racing offers a 280ZX factory turbo Boost Enhancement and Control kit (BECkit) which can virtually double factory turbo boost levels. The Stage I BECkit features air-to-air intercooling, turbo oiler, adjustable boost valve, boost-controlled auxiliary air flow system, safety valve and optional (Stage II) modified wastegate and exhaust system components. This system was decribed in a *Car & Driver* magazine article entitled "Maximized Maxima". The BSR Stage II turbo 280Z or ZX engine packages start at $4995.00 and will escalate depending on the final configuration and whether or not the system is to be installed by the people at BSR. The BECkits for factory 280 ZX turbos start at $795.00 (Stage I) and are designed for do-it-yourselfers and require 4–6 hours of installation time. Stage I 240, 260, & 280 (Z&ZX) bolt-on turbo systems are presently in the planning stages and will also be designed with the do-it-yourselfer in mind.

CALLAWAY

Media accolades are ubiquitously present in the Callaway Turbocharger Systems (CTS) information packet. These accolades address the layout and cosmetic features of the CTS turbo systems but by and large center on performance (superlatives and all); e.g., "Phenomenal acceleration" (Auto Week), and "amazing, near rocket material." (Auto Week). You too can join in on the fun and check out the super activities first-hand! CTS has undoubtedly done their homework when it comes to turbo systems designed for the performance nut. These systems are complete and use state-of-the-art controls and components such as wastegate boost control, intercooling, the Microfueler fuel-metering/enrichment device, and electronic knock sensing.

These systems sport Rotomaster T04b turbochargers carefully matched to the particular vehicle application. 100% horsepower gains are claimed for several applications. A 75% improvement in torque readings is attainable in the Stage II 1600cc VW gas engine version as a result of increasing the boost to 10 psi and adding the Microfueler, which is supplied with all BMW conversions and the Stage II VW conversion. The device works with Bosch K and L Jetronic fuel injection systems.

The Callaway plant in Old Lyme, Connecticut.

Mercedes' 240D is slow. Callaway's kit elevates its acceleration to decency.

Callaway's VW Scirocco 1.8 liter kit. Air conditioning is no sweat.

A top quality intercooled VW 1.8 liter kit is also available through Callaway. Make your GTI GIT!

Porsche is too busy building normally-aspirated 944s to bother selling intercooled 944 Turbos in the U.S., but Callaway would be pleased to sell 944 owners the kit of their dreams. IHI turbocharger puts out 10 psi boost. Combustion hambers and pistons are modified for low compression.

The Microfueler is a discrete fuel injection system which provides auxiliary fuel to the engine under boost conditions only. This fuel enrichment is necessary for fuel injected engines whenever high boost is being applied because stock fuel injection systems cannot usually meet fuel flow requirements. The Microfueler senses engine speed from an ignition signal and boost pressure via an intake manifold pressure sensing chip. On the basis of this information, the brain box precisely calculates the additional fuel needed and sends a variable pulse to the injector, instructing it as to how much fuel it should spray in. Various size Microfueler nozzles are available so the system can be precisely matched to engine requirements.

Callaway kits also feature ductile iron exhaust manifolds and 18 gauge stainless exhaust system components. Kits are offered for watercooled VW models, Porsche 944 and 928, Audi, BMW, Mercedes, and Saab vehicles.

Kit prices range from $1761.00 for the Stage I VW to $18,500 for the 928 Twin Turbo factory installed. In most conversions the stock fuel injection system and manifold are retained but are augmented by the Microfueler. The express warranty covers, essentially, defects in materials and workmanship for a period of 90 days or 3000 miles. All Callaway kits are "designed to be installed by someone who is willing to learn the ins and outs of automobile service procedures and who is willing to take the time to invest in the tools and the literature to do it properly." The instruction manual was designed to accommodate the first-time installer and up to two days of down-time can be expected for the vehicle.

In keeping with a desire to offer premium performance to the many enthusiasts who may be purchasing OEM turbocharged cars, Callaway offers the Bump-Up™ System. The heart of this package is a ram-air cooled intercooler, which provides substantially increased performance via cooler, denser intake air. This cooling of the intake air permits substantial increases in performance with very minimal increases in engine pressure and thermal stress, while also minimizing the tendency to detonate. Bump-Up System components typically include the intercooler and associated piping, as well as a Microfueler to increase the potential output of both turbo and engine. Options in-

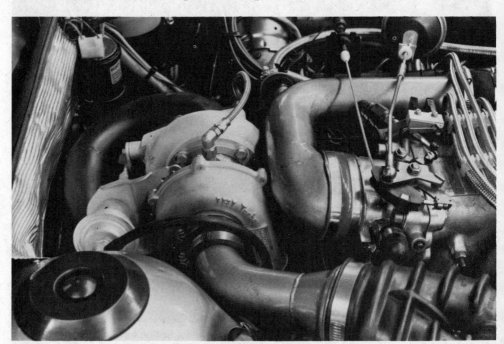

Audi's 1.7 liter Coupe is sweet, but Callaway's IHI turbo kit makes it crank.

Callaway's BMW 633 kit. How do you spell quality? Kit boasts boost-activated fuel enrichment, intercooling, and ignition retard.

BMW 320i turbo by Callaway. Clean! 7.3 second 0–60 mph acceleration is possible.

clude an enlarged exhaust system, cockpit boost control, and detonation sensing. The Bump-Up System increases the power rating of a Saab APC Turbo from 130 to nearly 180.

The general philosophy of Callaway Turbosystems is to offer high quality, complete systems that provide substantial performance increases within conservative engine stress limits. Factory installed systems feature combustion chamber modifications that lower compression ratio to the point where significant power increases are available with only minimal effect on engine stress. The overall approach is to offer a system designed primarily with regard to its performance and preservation of engine life, with price as a secondary consideration.

CARTECH

Cartech is one of a very few turbo kit manufacturers to offer the potential turbo kit customer a "Turbo Comparison Chart" for comparative evaluation. This chart encourages the potential customer to check out the competition by assigning points to each maker in response to 32 questions, total up the points and buy the winner. This challenge appears well-founded as Cartech does offer state-of-the-art turbo system design including the newest of the "hot-ticket" concepts in turbocharging—the variable area turbine nozzle. This concept virtually eliminates turbo lag through the use of variable vanes in the turbine housing that effectively control turbine speed. What the system does is to keep turbine speed up to an almost constant rpm under cruise throttle conditions, when turbo rpm ordinarily drops way off.

The turbo is an Aerodyne unit featuring self-lubricating ball bearings as opposed to the hydro-dynamic plain bearings used in most other turbos. A joint "technical exer-

Cartech's kit for Datsun's 8-sparkplug 4-cylinder NAPS-Z motor, used in the '80–83 200SX, retains fuel injection. The kit features a custom cast iron exhaust manifold and a boost pressure-activated ignition retard/fuel pressure regulation system.

Retain BMW quality and engineering with Cartech's 320i kit and get a BMW with boost. A twin turbo kit is available for the 633/733.

cise" between Cartech and Aerodyne produced a 1981 GS750E Suzuki motorcycle tricked-out with this new turbo system that went out on the track and proceeded to rein-in 130 horsepower.

Cartech also offers this new technology in 4-wheel form neatly packaged under the hood of a Toyota Supra—intercooled and all! Speaking of 4-wheelers, Cartech's main product line is 4-wheelers which include the marques Datsun, Mazda, Toyota, Mitsubishi, BMW, VW and Mercedes. Carbureted and fuel injected turbo systems are available and feature Rajay turbos (or, occasionally, an IHI unit) mounted on Cartech-designed cast iron exhaust manifolds. Boost—pressure ignition retard, cast aluminum, water heated carburetor mounting surfaces, braided oil lines, Edelbrock Varijection water injection, Holley (or occasionally Weber) carburetion, wastegate boost control and the new air-to-cool water intercoolers are a few of the other features offered with Cartech systems depending on the particular system and application.

Cartech's Mazda RX-7 kit uses an IHI turbo, a water heated carb mount, cast manifolds, a Holley 390 cfm 4-barrel, exhaust gas temperature and boost gauges, and a K&N air filter. This kit brings horsepower up to 140 and an intercooled system is in the works.

The Datsun 280 kit comes with an IHI turbo, boost gauge, cast manifolds, boost-activated ignition retard, and a fuel injection pressure regulator which increases fuel system pressure linearly as air flow increases. The 240/260 Z kit uses a Holley 390 cfm 4-barrel, K&N air filter, Edelbrock Varijection water injection, a water-jacketed carb-mount, and boost gauge.

Two kits are available for the 200 SX. The '80–'83 NAPS-Z kit retains the fuel injection and comes with boost-activated ignition retard and a fuel injection regulator which increases fuel pressure as air flow increases. The other 200 SX kit, which is also avail-

Cartech's complete kit for the DOHC Supra includes a carefully designed exhaust manifold, fuel injection pressure regulator, Bosch fuel pump, and boost pressure-activated injection retard for the '82 model. Intercooling is optional.

able for the 510, 610, 710, and pickup, uses a Holley-Weber 2-barrel carb, K&N air filter, and a water heated carb mount.

The DOHC Toyota Supra kit is good for 220 horsepower and 240 with the optional intercooler. 6.5 second 0–60s are possible with the base kit, which uses a Rajay turbo, linear fuel pressure regulator, Bosch 100 psi fuel pump, boost gauge, cast manifolds, and a boost-activated ignition retard system for the '82 model. Toyota 20R and Mit-

The Cartech Datsun 280Z/ZX produces boost at a low 2000 rpm, thanks to an IHI turbo.

subishi 2.6 liter kits are also available. A twin turbo intercooled BMW six cylinder kit puts out 310 horsepower in 3.3 liter trim. Rotomaster turbos are used with no detonation at 10 psi on 91 octane gasoline.

Cartech has also entered the diesel pickup market which its new Toyota Diesel pickup turbo kit which claims a reduction in 0–40 times of 4 seconds and a fuel economy improvement of 3 to 5%. Kit installation times will naturally vary based on the particular kit, but in the case of the Toyota diesel pickup, Cartech says you will be looking at 8 hours installation time. Suggested list prices on Cartech basic kits start as low as $1225.00 (Toyota Diesel) and spiral up to $4550.00 (BMW).

Performance will also vary depending on application. However, the Datsun 280Z/ZX kit does let the turbo cat out of the bag, posting 0–60 acceleration times of 6.3 seconds and a top speed in excess of 135 mph! Cartech kits are offered complete including the vital instructions for the do-it-yourselfer and component parts whose quality is second to none. An additional feature of most Cartech systems is that a six month component warranty is offered—just in case! Custom and special application turbo system designs are offered by Cartech along with individual components for those wishing to design their own systems. To top it all off, Corky Bell, Cartech's owner and chief engineer is a graduate engineer and is most willing to offer advice and help if so requested.

IPD

The Volvo specialists at IPD have designed a turbo kit for the those 242, 244 and 245 models which use the B21 engine. Installation is designed for the do-it-yourselfer and should take less than six hours, according to IPD. Once the system is installed and checked out, you should expect to see a power increase in the range of 30–40%, with water injection preventing knock and 6 psi of boost pressure driving home the fuel/air mixture. IPD has also experienced a small increase in gas mileage with their own turbo Volvo. A new 240 series fuel injection turbo kit for the B18/B20 engines is also avail-

IPD's kit fits B211 engines in the common 240 series Volvos. This a low pressure system which uses the stock fuel injection.

The IPD Volvo kit is designed for easy installation.

able. While the turbo kits are not presently approved by the Oregon Department of Environmental Quality nor on the "exempt list" of the California Air Resources Board the matter's being pursued with high hopes of getting both sanctions in the near future. The turbo Volvo kit received some national recognition in the July '78 issue of *Road and Track* magazine. The B21 kit was designed to cost $1600–$2000 over-the-counter with the customer option of having IPD install the system at an additional cost.

JAFCO

Fans of fuel injected Alfa-Romeos can stand up and be counted in the "turbo-mania" movement and Jafco turbo systems has made it all possible. Tests were made on a 1974 Berlina 2000 with a Jafco system installed and all else virtually stock. The engine cranked out 150 horses at 6000 rpm and 0–60 times of 7.6 seconds while the turbo generated a maximum boost pressure of 11 psi. Boost pressure is controlled with a Boost-Guard wastegate system while the fuel enrichment is handled by Jafco's High Speed Fuel Enrichment System set to enrichen at 5 psi. A cast, heat resistant alloy exhaust manifold supports the AiResearch T04 turbocharger, while the fabricated induction tubing gets the charged air to the valves. The whole Jafco system is neat and to the point and looks like a factory installation. Three basic turbo kits are available covering the 1975–79 Alfetta Sprints and Sedans, the 1971–74 1750–2000 models and 1975–79 Spiders. Water injection, TRW Forged turbo pistons, CD (capacitive-discharge) ignition system and several different turbine housing sizes are options. These deluxe kits are priced at $1850.00 and are supported by a six month limited warranty on materials and workmanship.

LARSON

The Larson Corvette single turbo system has been around since 1977 featuring blow-through design with a Holley carburetor and a single AirResearch T04 turbo, a flat-hood installation, and stainless steel exhaust piping. A Larson-designed adjustable pop-

A Roto-Master turbocharger pressurizes DOHC Alfa-Romeos in this Jafco kit. System uses a Boost-guard wategate and enrichens the fuel injection above 5 psi boost.

off valve in the cast aluminum air bonnet provides boost control, which is set in the 12 psi range for the L-48 and L82 if nourished on 91 octane, but can be moved up to 15 psi on 104 octane and an incredible 20 psi on Cam-2 racing gas (with 0-ringed heads and forged pistons). An automatic-transmission equipped Vette with a Larson turbo system has rushed through the quarter mile in 11.8 seconds.

The Larson turbo kit includes everything needed to do the job including the water injection and distributor retard kits. Air conditioning and cruise control can be retained with Larson turbo kits which is good news to those appreciative of such amenities. The Larson twin turbo kit is of blow-through design featuring twin IHI turbos, an Edelbrock 2101 intake manifold and Holley carburetion. The system fits under the stock hood——air filters and all. The '82–84 Corvette ported fuel injection is usually discarded in favor of the Holley carb set-up for the twin turbo installation. A wastegate system is offered as an option on the twin turbo and is located in the crossover pipe. Exhaust pipes are constructed of 2 ½″ diameter stainless steel tubing and will hook-up to the 1973 dual pipes or the Larson side pipe option ($1000). The Larson basic kit

Larson's small block Vette kit is a show-stopper. This one is a single turbo blow-through system using an AiResearch TO4 turbocharger, custom exhaust manifold, water injection, and ignition retard.

Larson's twin-turbo small block Vette kit uses IHI turbos. It's tight, but even the air filters fit under the hood. Turbo blow-through Holley carburetor.

starts at $2000.00 for the single turbo system and $5000.00 for the twin turbo set-up. Attention to cosmetics and ease of installation are surely the icing on the cake with the two Larson turbo systems.

MARTIN

Martin Turbo is dedicated to the enthusiast, however, shear gut-raking power is not the prime objective even though their systems do perform. They design turbo systems which use stock carburetion and ignition systems making installation easy and factory-like in appearance. In this day of emission regulations, Martin has come to the rescue by offering several systems meeting California Air Resources Board (CARB) requirements for street and highway use. This is undoubtedly a plus if you are considering turbocharging your street vehicle and concerned about the legalities of it all. Martin, while heavily into GM 305 and 350 turbo systems, also offers a VW Baja and Sand-buggy turbo system, a dual turbo marine system and several Chrysler turbo systems. Specific applications in the GM line include; Suburban, Blazer, Van, Camaro and Corvette (305 and 350's only), and the Chrysler line includes the Dodge Pickup (2 and 4-WD) and Dodge/Plymouth passenger cars (318, 340 and 360).

The systems are usually offered in Stage I and Stage II trim with maximum boost selection of 5 or 9 psi. The kits are complete bolt-on systems for ease of installation

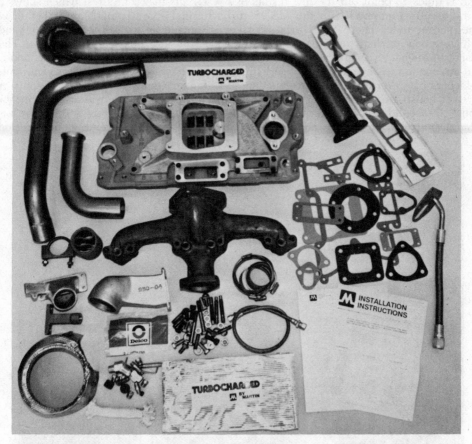

Martin small block Chevy kits are designed for specific years and models. You selected the turbo size and options—don't forget the vital wastegate.

and are backed by a one year limited warranty covering material and workmanship. A draw-through turbo design is used, featuring a Martin designed "polarity valve" which is claimed to reduce turbo-lag by allowing incoming air to by-pass the turbo under conditions of high intake vacuum. Turbo speed usually drops off quite a bit under high vacuum. Bypassing airflow has the effect of increasing turbo rpm over what it would be otherwise.

Wastegates are recommended for heavy duty applications. Otherwise the systems use internal boost control valves, which limit the intake manifold pressure to some predetermined level (usually 5,7 or 9 psi). Water injection and intercooling are available as is a Stage I upgrade kit which effectively makes a Stage II system out of your Stage I. Rajay turbochargers are the featured unit in Martin turbo systems with several turbine housings to choose from for dialing in the turbo to specific applications. OEM turbocharger performance improvement kits are also available which Martin claims can increase rear wheel horses by 32%. These kits will fit the Buick 3.8 turbo (1977–1980), Pontiac 4.9 turbo (1980) and Ford 2.3 turbo (1979–1980) and come complete with water injection, boost gauge, mounting panel and all necessary hardware.

MARTIN SCHNEIDER

Martin Schneider Designed Systems (MSDS) specializes in 914 Porsche turbo systems and is quick to point out the test results of *VW Greats* magazine's test which shows a two-liter 914 equipped with the MSDS turbo turning 15.49 seconds and 90 mph in the quarter mile. The MSDS turbo kit takes advantage of the 914 stock power band which comes on like turbo gangbusters at 3000 rpm and holds up to red-line. The Rajay turbocharger is specially modified by MSDS and pushes 7 psi worth of boost. An automatic fuel injection boost–enrichment module guards against mixture lean-out, which can be detrimental to the function and life expectancy of any turbo engine. A pre-set boost controller is used to guard against over-boost while you are looking for those 15 second quarter mile times. Installation requires no welding and can be accomplished with hand tools and eight hours to "less than a weekend" depending on

Martin Schneider Designed Systems Porsche 914 2-liter, viewed from underneath.

MSDS Porsche 914 2-liter kit viewed from the top.

your wrench-pulling experience. An added feature of this system is that it "doesn't say goodbye to your 914's stingy fuel economy." The MSDS turbo also allows you to continue using the 914's heater system, as the heat exchanger is left in stock position.

If you are so inclined, MSDS can also fix you up with a 914/911 turbo terror. Yes, they will drop a 911 (3.3 CIS–injected engine) into your 914 complete with a turbo system built out of stainless steel and heavy gauge pipe plus wastegate control—and water injection to boot! This combination was featured in *"VW & Porsche"* magazine, which claimed a 0–60 performance of less than 5 seconds! Not to be left out of the diesel turbocharging market, MSDS has designed a diesel VW Rabbit/Jetta/Pickup turbo system featuring an IHI turbo with integral wastegate. This nine psi system will give you a 50% increase in power, reduced smoking and improved fuel economy. A "breather assist" (reduces oil blow-by), aircraft quality plumbing, complete exhaust system, and fuel enrichment control are standard features of this system. Optional goodies include a diesel tachometer (VDO), turbo boost gauge, Carrera GT hood scoop and cast aluminum intake manifold. The diesel turbo system tips the dollar scales around $1195.00 for the basic kit. MSDS offers other accessories in their product line-up including an exhaust wastegate with remote boost control and a boost pressure gauge for the Porsche 924 Turbo.

PFAFF TURBO

Pfaff offers turbo kits for the Chevy small block V-8, BMW 4 and 6 cylinder in–line engines, and the iron duke Fiero. All Pfaff kits are blow-through systems. A twin turbo Chevy small block "builder's" kit includes two IHI turbochargers, cast iron exhaust manifolds, a Holley Street Dominator aluminum intake manifold, a Holley four-barrel carburetor, modified by Pfaff for blow-through operation, two K&N air filters, a boost gauge, a cast aluminum carburetor hood (Gale Banks-style, but unpolished), Holley mechanical and electrical fuel pumps, an MSD 6A ignition module with boost–controlled retard, and hoses, oil lines, tubing, etc. This system requires some fitting and machining for various displacements and engine compartments.

A separate twin turbo Chevy small block kit is available for '82 and later Camaros and Firebirds with the 160 horsepower LG4 305. It will not fit the crossfire injected or 190 horsepower L69 305s. What makes this kit unique is an intercooler. Also included are two IHI turbochargers, a polished carburetor hood, MSD 6A and Turbo Master ignition control, twin K&N air cleaners, Holley electric fuel pump, and chromed intake tubing. This kit will turn a leisurely Camaro/Firebird into a screamer capable of 6.2 second 0–60s. But Pfaff warns that brakes, tires, transmission, and rear axle should be upgraded to handle the speed and power.

The Fiero kit drops 0–60 mph acceleration from 11.6 seconds to a claimed 7.9 seconds. Pfaff includes a 24 month/24,000 mile drivetrain service contract with this kit. A maximum boost of 7 psi is permitted. An IHI turbocharger, boost gauge, cast iron exhaust manifold, cast aluminum intake hood, ignition retard and fuel enrichment—under—boost system, and chromed intake tubing are included.

Pfaff's twin turbo Chevy small block "builder's" kit includes cast exhaust manifolds, IHI turbos, aluminum carb hood and intake manifold, Holley carb, fuel pumps, ignition retard system, and more. Some fitting and machining is necessary.

Pfaff Stage 2 BMW four-cylinder kit blows through two Mikuni sidedraft carburetors.

Pfaff has the BMW line well covered. The four-cylinder kits cover the 2002, 318i, and 320i. Stage 1, 2, and 3 kits are available for each. The stage 1 kit offers an IHI turbo, cast iron exhaust manifold, boost gauge, fuel pump, a Weber DGV downdraft carburetor and cast aluminum carb hood for the 2002 only, and a Pfaff fuel enrichment system for the 318i and 320i. The stage 2 kits switch to two Mikuni—Solex sidedraft carburetors with an aluminum intake manifold on the 2002 and 320i, and add an adjustable ignition retard system and turbo head gasket on all three models. Stage 3 tops the Stage 2 system with an intercooler, driver-adjustable boost control, modified profile camshaft, 6.9:1 compression pistons, and an exhaust gas temperature gauge.

The Pfaff has BMW 6-cylinder twin-turbo kits for the 528i, 530i, 533i, 630i, 633i, 635i, 733i, and 745i. Again, three stages are offered. Stage 1 includes two IHI turbochargers, cast iron exhaust manifolds, cast aluminum intake tubing, boost gauge, air filter, Pfaff fuel enrichment system for fuel injected models, and water injection. Stage 2 adds a turbo head gasket and an adjustable ignition retard system. Stage 3 kits top the bill with an intercooler, driver-adjustable boost control, 7.2:1 compression forged pistons, camshaft, dual disc clutch with aluminum flywheel, and exhaust temperature gauge.

Pfaff also sells a complete turbo-kitted double overhead cam four cylinder Toyota engine which bolts into any '70 to '84 Celica, Corona, or Toyota pick-up. Pfaff imports these 2-liter engines and rebuilds and modifies them for turbocharging with 7.5:1 compression pistons and block o-rings. The turbo system features an IHI turbocharger, cast iron exhaust manifold, two Mikuni sidedraft carburetors with cast aluminum carb hoods, fuel pump, air filter, adjustable ignition retard system, driver-adjustable boost control, exhaust gas temperature and boost gauges, and a heavy duty clutch. The intercooler (included) requires special mounting depending on the car, and that problem is left up to the installer. A custom exhaust system is also required. Drop one of these engines into your four-cylinder Celica, and you'll make Supras the laughing stock of the highway.

SHANKLE

The Alfa-Romeo fuel injected 2000cc engine is turned into a real screamer by Shankle Automotive Engineering——a screamer that gives 160 bhp from an engine fitted with 7.8:1 compression ratio forged pistons, and a 45 DCOE Weber carburetor via 15 psi of boost generated by the Rajay turbocharger. A Rajay wastegate controls the boost levels and a Shankle water injection system prevents detonation. An oil cooler kit keeps the oil temperature at desirable levels while a special turbo exhaust system gets rid of the leftovers.

If it sounds like a truly complete kit, it is! And no wonder, as Shankle has been in

Shankle's kit for the 2-liter Alfa even includes forged low compression pistons. The stock fuel injection is canned for a Weber 45 DCOE carburetor with progressive linkage.

Nothing like a turbocharged double overhead cammer! Shankle's Alfa kit uses a whopping 15 psi boost and a boost pressure-activated water injection/ignition retard system.

the Alfa race and hi-performance business since 1967. Their race and competition knowledge and attention to details has been put to good use in developing this turbo kit. The kit is so complete that installation, of course, requires engine disassembly to install the forged 7.8:1 turbo pistons. The installation of an oil cooler and turbo muffler are also required. However, standard hand tools will get the job done with one exception—welding the exhaust pipe. The instructions are well written (27 pages) and are illustrated which is of vital importance to the do-it-yourselfer——and some professionals too! The additional effort expended during installation is long forgotten when you can accelerate that Alfa from 0–60 mph in 7 seconds!

SPEARCO

Spearco might very well be the granddaddy of aftermarket turbocharger kits by virtue of introducing their 2000cc Pinto kit in 1971. They have virtually led the turbo kit parade to its present high technology berth in the world of performance automotive products. In addition to developing and patenting the first electronically-controlled water injection system, Spearco also pioneered the use of wastegates and electronic controls on turbo systems.

Spearco uses both Rajay and IHI turbos in their 30 or more kits, depending on the application. The Spearco domestic vehicle kit line includes the Chevy big block and small block and Ford 351M, 400, Escort/EXP/Lynx/LN7, Pinto, Mustang II, Fairmont, Capri, and Courier. Import systems include the VW Rabbit (diesel and gas), pickup, Baja Bug, Scirrocco, and Beetle, Toyota's Celica, Pickup (diesel and gas), Supra Twin Cam, and Cressida Twin Cam, the Isuzu and Chevy LUV 4-wheel drive pick ups. These kits are for the most part offered as bolt-on systems with no fabrication involved, however, the universal small block Chevy turbo kit may require some exhaust system fabrication to meet a specific application.

Carburetion may or may not be part of the kit depending on customer wishes but in most cases any recommended optional carburetion is available through Spearco. The 351M–400 Ford kit, for example, will accommodate the stock 2-barrel carburetor but Spearco recommends and offers an optional 4-barrel Holley if you really want to do it right! The IHI turbo systems feature an integral wastegate while the Rajay systems may require an optional wastegate system.

Spearco offers an optional water injection system and can supply other individual turbo system components in case you want to design your own turbo system. Speaking of optional components, Spearco is offering an aluminum aircraft style intercooler, as part of their new Toyota Supra Stage II kit, which effectively ups the Stage I non-intercooled boost over 2 pounds with a more than proportionate performance increase. The Ford Escort/Lynx, VW Sedan and Toyota 4-wheel drive kits are of the blow-through

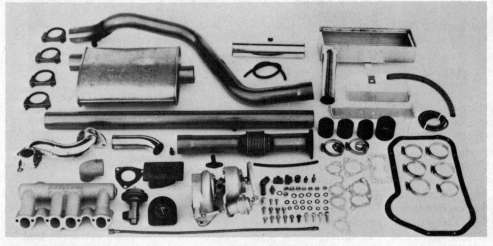

Spearco makes the diesel Rabbit hop to 60 mph in 12.1 seconds with an IHI turbo. A boost-activated diesel enrichment system and complete exhaust is included.

More pick-up for the Toyota 4WD pickup is available from Spearco with this blow-through turbo kit. Factory air conditioning and power steering are unaffected.

Spearco Chevy small block pick-up kit is complete, right down to the kick-down linkage for automatic transmissions.

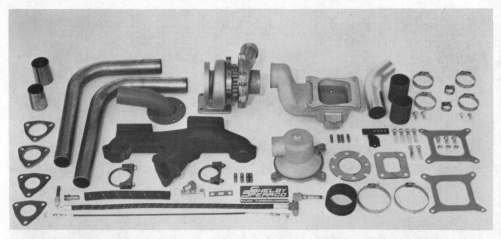

Spearco Chevy small block kit uses slope-flow carb adapter, and blow-through hood.

Blown off by a Corvair-powered buggy? Spearco sells this Baja Bug kit to boost the dual-port air-cooled VW. It won't fit under a Beetle's hood. Low compression pistons, intake manifold, and Holley/Weber 5200 2-barrel carb are necessary.

DOHC Supra kit by Spearco uses an IHI turbocharger and optional intercooler.

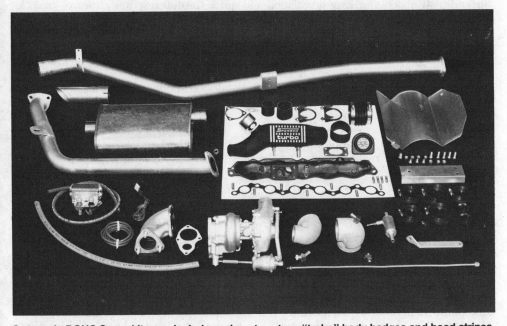

Spearco's DOHC Supra kit even includes exhaust system, "turbo" body badges and hood stripes.

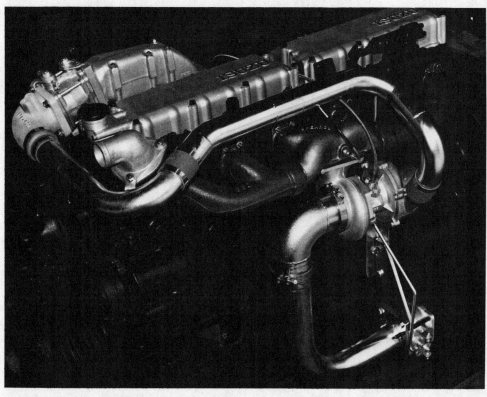

Spearco even has a kit for the 6-cylinder Isuzu diesel which retains the fuel injection.

turbo design making for easy installation and quick throttle response. Spearco basic kit prices start at $1095.00 (VW) and escalate to $2260.00 (Toyota Twin Cam).

Performance with Spearco systems varies from application to application. The Toyota Supra Stage I kit will improve 0–60 times by 1.3 seconds (Stage II is better yet) with only 5 psi boost, while Rabbit 0–60 times show a 3.6 second improvement with 7 psi boost. The Toyota Supra 2.8 Twin Cam and 1983 Cressida versions are certified to EPA standards and have pending CARB exemption status in California. This Toyota twin cam system uses the stock oxygen sensor, and a boost-activated ignition retard/fuel injection enrichment system.

Spearco expertise is very widespread. Large diesel engines used in heavy construction equipment and bus applications overseas such as the Isuzu E-120 and 6BOL diesel engines are the newest marketing venture for them. An intercooled turbo system for marine use on the Isuzu UME-120 engine is also in the offing.

TURBO INTERNATIONAL

Despite its name, Turbo International concentrates on domestic vehicles. Their kit line-up includes the Dodge/Chrysler 440 engine (popular in musclecars and motorhomes), Chevy small block (one low hoodline kit for Corvettes, Camaros, etc, and another kit for pickups and other light trucks), and a Chevy big block kit. These systems are complete and offer everything from air cleaner and catalytic adapter pipe to safety wire and polished aluminum components. An air conditioning option is available for the Corvette. "No add-ons, no modifications, no hassles!" is what the literature says. Water

Turbo International's Chevy small block kit is good for a 50% power increase with the water injection that's included.

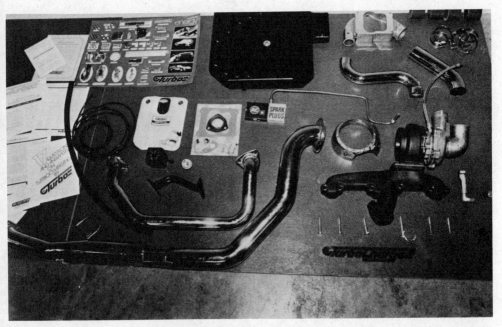

Turbo International's kits are simple, well-designed, and complete. They are also sold by Corvette America.

injection reduces detonation, while a specially designed exhaust expansion chamber reduces turbo back pressure by incorporating an enlarged area immediately behind the turbo discharge. Show quality chrome is used in liberal amounts to accentuate the sanitary, draw-through turbo system plumbing.

Provisions will have to be made for the air pump in the emission control system either through relocation or total removal. Stock carburetion is maintained, although the PCV vacuum tube will have to be replaced with a new one supplied in the kit. This kit does not require removing the oil pan from the engine for the turbo drain-back line but rather incorporates the turbo drain-back fitting into a modified fuel pump mounting plate. Installation instructions are complete and well-presented. The literature suggests that you should plan for 6–10 hours of installation time. Only standard hand tools are required. A troubleshooting guide is also included in case one has not followed the instructions quite closely enough.

The Corvette kit is also available through Corvette America. The Dodge 440 kit has received CARB approval for model years up to 1979, and the Dodge small block kits are approved for automatic transmissions up to 1981. A 50% power increase can be expected with either kit.

TURBOCHARGER, INC.

Turbocharger, Inc. entered the diesel car turbo kit market years ago after specializing for many years in turbocharging diesel trucks, marine engines, and heavy duty construction equipment. Their first automotive assignment was to turbocharge a Mercedes Diesel which was quickly followed by a Peugeot 504 diesel kit. The success of the Mercedes system is epitomized in the words of the people at *Road & Track* magazine who said "anytime you knock nearly 10 seconds off a 0–60 (mph) time you've got something." Add to this obvious power gain a reduction in diesel smoke (particulates re-

Turbocharger, Inc.'s kit fits 200Ds, 220Ds and early 240Ds.

duced approximately 75%) and an increase in fuel economy (12% in town and 10% highway) and it appears that Turbocharger, Inc. is doing something right!

These kits are no longer in production but Turbocharger Inc. still has them in stock. The systems for the Mercedes and Peugot are complete up to the exhaust system and are backed by a one year or 24,000 miles warranty against mechanical defects. It is not necessary but *is* advised by Turbocharger Inc., that you have a qualified mechanic install your system. An 11 hour installation time is claimed. Turbocharger, Inc. points out that while their turbo diesel systems have been tested by independent testing facilities and shown not to "place an unusual or burdensome strain upon the engine or drivetrain," the turbo systems may still invalidate any vehicle manufacturer's warranties. In addition to producing and marketing the above-mentioned kits, Turbocharger Inc. has complete repair and service facilities for the repair of turbochargers and is also a distributor for Schwitzer, Rajay, and IHI. This is a reassuring bit of information for those who weigh after-sale service into the decision making process when it comes time to purchase any product.

TURBO TOM'S

High-output turbocharging is Turbo Tom's specialty especially when it comes to Datsun (Nissan) Z-cars. Turbo Tom's Datsun turbo system line-up covers the Z and ZX, Maxima, and L16/L18/L20B engines. Kit prices run from $2585.00 (4 cylinder) to $2985.00 (ZX) complete and ready to install. And to top if off, Turbo Tom's claims to have "the world's most detailed instruction manual" for do-it-yourselfers. A Rotomaster T04B turbocharger custom—matched to the application heads up the standard system components list followed by the TT designed, heavy-wall ductile iron exhaust manifold and cast aluminum intake manifold. The 390 CFM (optional 600 CFM) Holley carb is subjected to 13 modifications to improve it for the task at hand. And just to make sure

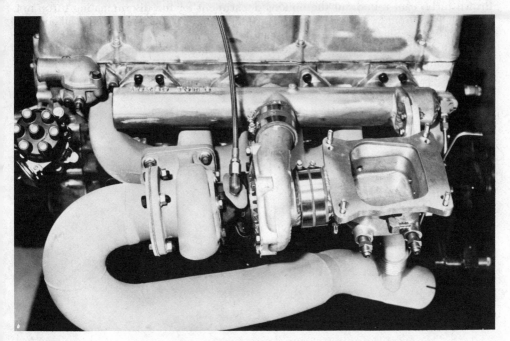

Turbo Tom's Z-car kits use a Holley 4-barrel carburetor. Nitrous oxide is an option.

Turbo Tom can make 4-cylinder Datsuns haul, too.

the modified Holley does not run dry in the heat of battle, a Carter vane-style electric fuel pump along with Aeroquip fuel lines is also supplied in the standard kit.

A wastegate complete with a 2 psi adjustment capability, high-output water injection, Dial-A-Boost boost adjustment kit, Holley 600 CFM double-pumper and a Racer Brown turbo cam round out the options department for the discriminating turbo nut. And, oh yes! Would you believe nitrous oxide too? The Z Stage I system will accept 9 psi of boost, the Stage II Z system can handle 15 psi and the super trick, "built" engine can produce 21 psi on the gauge. Complete kits and/or component parts can be purchased from Turbo Tom's; however, the turbo kit line at present is pretty much dedicated to the Datsun marque.

6
Three Step by Step Installations

The following step-by-step turbo kit installations demonstrate in the most graphic manner exactly what is involved in installing an aftermarket turbo kit. To cover all the bases, an in-line engine, V-8, and rotary are all included, as are both carbureted and fuel injected engines. We've even got an air—to—water intercooled kit for plumbing lovers.

Two of these kits, the CarTech RX-7 and the Arkay Toyota Supra, are designed for a specific engine in a specific car, and hence are complete right down to the hose clamps, requiring a minimum of custom fitting. The Ak Miller kit installed on the Ford Bronco, on the other hand, is intended to provide the major components needed to turbocharge a wide variety of Ford 351 Windsor-powered cars and trucks. So this kit requires welding, sheet metal cutting, the purchase and installation of a different ignition system, a willingness to "gofer" hardware, a problem-solving frame of mind, and, even more important, a thorough understanding of engines and turbocharging.

ARKAY'S TOYOTA SUPRA IN-LINE SIX CYLINDER TURBO KIT INSTALLATION

1. First step is to fill-up with premium gasoline and disconnect the battery cables. Jack up the Supra and remove the chassis undershield to expose the engine bay. The exhaust pipe between the manifold and catalytic converter, as well as the converter itself, are removed. If you're installing the optional exhaust system, remove the entire stock exhaust system.

The engine oil is drained and the oil filter replaced. The charcoal canister is removed. At the rear of the cylinder head is a pair of heater hoses joined by a connector pipe. The hose that is connected to the engine and the connector itself are removed. An inch is cut off the top hose and it is hoseclamped back onto the engine so that the water circuit is again complete.

The windshield washer bottle is temporarily moved out of the way for working room. The oil pressure sending switch (just to the left of the oil filter) is replaced with a T-fitting, and the stock intake crossover pipe's hoses and air canister are removed.

2. The exhaust manifold adapter and turbocharger are installed as a unit. They are slid down between the engine and inner fender from above——it's a tight squeeze, but the assembly WILL fit.

3. An adapter connects the turbocharger's compressor discharge to the Supra's stock

intake crossover pipe. An additional support bracket is installed between the exhaust manifold adapter and the manifold, so the lower manifold studs don't carry the full load. Now all the nuts are final tightened.

4. The turbocharger drain hose and fittings are connected first to the bottom of

the turbocharger, then to the engine oil pan. This banjo fitting and "flow-through" bolt replace the standard oil pan drain bolt.

  5. The oil pressure hose is connected between the top of the turbocharger and the T-fitting installed in the engine earlier.

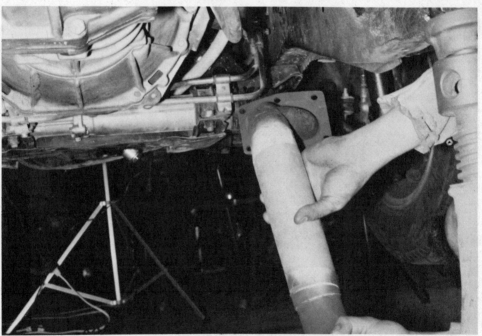

6. The exit pipe is bolted to the bottom of the turbocharger. If you're using the stock exhaust system, you now fit the short section of included straight pipe between the exit pipe and stock exhaust pipe. Otherwise the optional Arkay large diameter exhaust is now bolted to the stock hangers and exit pipe.

7. The turbine heatshield fits behind the top of the exit pipe. Clearances are tight

here, so a little patience pays off. The shield actually bolts to a charcoal canister mounting hole in the firewall and the windshield washer bottle frame. A hole must be drilled in the washer bottle frame, and a little heatshield bending is often necessary. Another heatshield protects the underhood insulation blanket and is attached directly to this insulation blanket.

8. The turbo intake pipe is connected between the turbocharger inlet and the stock air flow meter's rubber hose.

9. The air vent hose which was connected to the throttle plate casting is now routed to a fitting on the turbo intake pipe (just installed). The opening in the throttle plate casting is plugged with a rubber cap and clamp.

Behind the intake manifold is a fuel-retaining banjo bolt which must be removed. A hole is drilled through the banjo bolt's head and it is tapped. After the bolt is reinstalled, an elbow fitting is threaded into the hole for the auxiliary fuel injector. The fuel line is routed from this elbow fitting to the auxiliary injector in the previously installed turbo intake pipe. Install the tie-wraps to secure the hose; this will protect the fuel line from heat. This must be done very carefully because a melted fuel line causes a hell of a fire!

10. The Arkay ADC/AFM (ignition retard/fuel enrichment system) control box is mounted on the firewall. The black wire from the control box is grounded to one of the box's mounting screws. The control box's red wire goes to a Supra hot lead. Two other control box wires go to the Supra's stock ignition, under the ignition coil. A long wiring harness is then routed from the control box to the auxiliary injector.

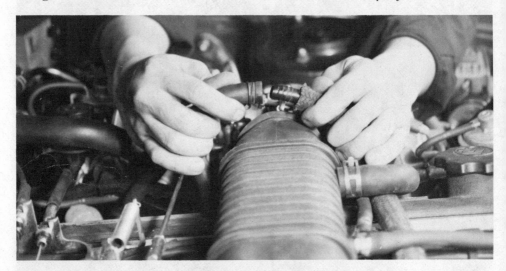

11. A cruise control vacuum line is tapped with two T-fittings, one for the ADC/AFM control box, and one for the boost gauge. One hose is hooked up between a T-fitting and the control box. The boost gauge hose is routed through a stock firewall grommet to the boost gauge, which is installed on the dash. Light wires for the gauge must be hooked into the light switch or dash light wires and a ground.

12. The engine is filled with oil and the radiator coolant is topped off. If you plan to use high boost pressure frequently, Arkay recommends you install colder spark plugs. After preliminary ADC/AFM control box settings are made, it's time for a test drive. Boost pressure, fuel enrichment, and ignition retard adjustments can now be made. After a few hours check all the hardware and hose clamps for tightness.

AK MILLER'S FORD BRONCO V-8 TURBO KIT INSTALLATION

1. It's this pile of discarded emissions equipment that makes the installation of a turbo (except a few approved kits) on any vehicle illegal in California except for off-road use. The restrictive stock exhaust and miles of smog hoses on this '83 Bronco would severely hamper turbo performance if left in place.

2. It takes almost a day's labor to remove all of the extraneous hoses to get down to the actual engine, in this case a 351 Windsor, which should have gobs of torque for off-roading with the Ak Miller turbo set-up.

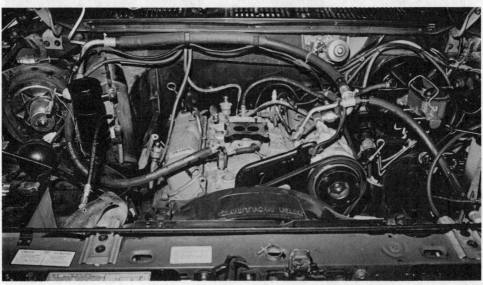

3. The stock driver's side exhaust manifold is modified with the addition of this tubing and plate to accept the AiResearch turbocharger. Welding on a used manifold isn't recommended for a turbo installation, as the heat will eventually crack it. Start with a brand-new manifold with no carbon inside. With proper welding with a Ni-rod arc, it'll last forever.

4. The new, modified driver's side exhaust manifold is bolted to the engine to begin the turbo conversion.

5. The AiResearch TO-4 turbocharger, an adjustable wastegate, and the Ak Miller click-stop for setting the wastegate limit. The click-stop can be used either on the engine or inside the vehicle.

6. With the turbo bolted to the manifold, the compressor housing bolts are loosened, allowing the discharge to be swiveled for the shortest plumbing path to the carb.

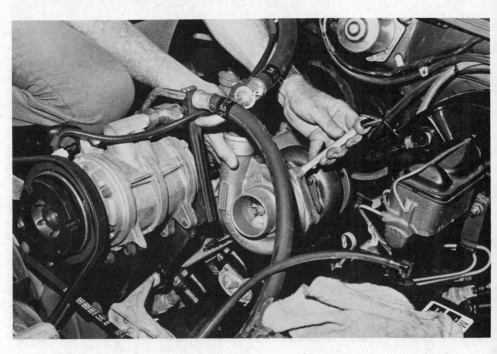

7. The stock '83 electronically-controlled carburetor (left) must be replaced with an earlier two-barrel Motorcraft carb to work with the turbocharger as a blow-through setup. Throttle linkage, vacuum hoses, and fuel line modifications are all necessary. Later, test driving will be necessary to select the optimum carburetor jet sizes.

8. An Ak Miller bonnet is bolted to the carburetor and a curved section of pipe (arrow) is used to connect the compressor discharge to the hood, using high-temperature hoses.

9. Turbocharged engines need plenty of fuel under extra pressure when under boost conditions, so the stock Ford fuel pump (left) must be replaced with a Carter high-volume pump.

10. In this blow-through installation, the fuel pump must have a balance line above the diaphragm. The Carter pump has been drilled and a fitting screwed in (arrow) to connect with a pressurized above-the-throttle source. This increases fuel pump pressure linearly with increases in boost pressure.

11. With the number of variations in stock vehicles, Ak Miller believes it's almost impossible to have a "turbo kit" for every car model. Many pieces have to be fabricated to fit the particular installation, like the exhaust crossover pipe which connects the exhaust manifolds.

12. The wastegate mounts under the Bronco, since it must be connected to both the crossover pipe, where it receives exhaust pressure and relieves it, and to the exhaust pipe, where it dumps the excess exhaust when the engine hits the boost limit.

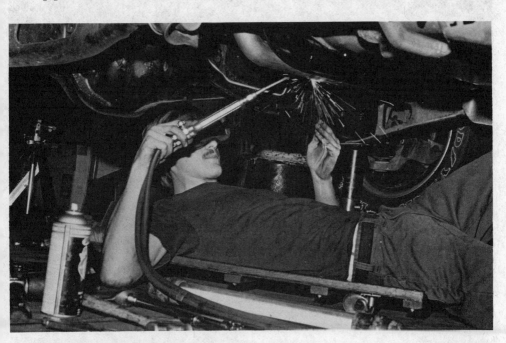

13. Just behind the Ford air-conditioning compressor lies the turbocharger. An Ak Miller cold air induction air cleaner has been adapted to fit directly onto the compressor inlet with a curved three-inch high-temperature hose.

14. The Bronco has two batteries and part of the left battery box must be trimmed away to clear the air duct hose that connects the air cleaner to a cold air duct hole behind the grille (arrow).

15. One of the factors that makes a turbo installation expensive on an '83 Bronco is that all of the computer-controlled parts must be replaced. The distributor at left has no advance mechanism at all, so it must be replaced with an earlier distributor and the correct Duraspark ignition module.

16. Fresh oil for the turbo is supplied with a line from the oil pressure sender fitting. The oil drainback must be installed above the oil level of the pan (arrow), and the hose must be routed without dips or kinks.

17. Because the turbocharger increases underhood temperatures, the stock thermostat at left must be replaced with a 160° unit.

18. A completely new exhaust system is fabricated from 2½ inch tubing with as few bends as possible, using an Ak Miller turbo muffler, which is free-flowing yet quiet enough for street use with the turbo.

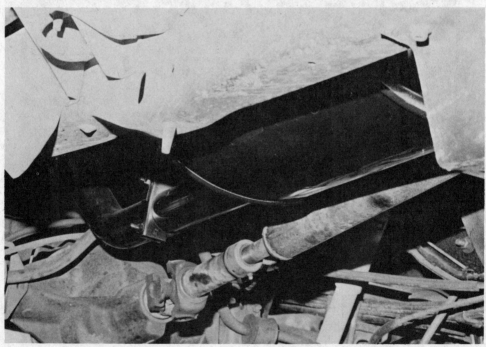

19. Even with the addition of the turbo, carb bonnet and new air cleaner, the finished Bronco installation is actually less cluttered than stock. Fresh oil and coolant is added, and ignition timing, boost pressure, and carburetor adjustments are made.

CARTECH'S MAZDA RX-7 ROTARY TURBO KIT INSTALLATION

1. The RX-7 is first filled up with premium gasoline, and the engine oil and coolant are drained. All these items have to come off: the stock air filter, carburetor, intake manifold, catalytic converter, air pump, exhaust system, vacuum valves and related plumbing, oil pan, hot start assist oil metering valve pushrod, fuel pumps, windshield washer bottle, radiator overflow tank and mount bracket. Wow! All the wires no longer needed are removed from the wiring harness. With the oil pan off, the oil drain fitting is installed, requiring a three-quarter inch drill bit. The oil pan is then reinstalled.

2. The oil inlet and drain fittings are screwed into the turbocharger, which is then bolted to the CarTech exhaust manifold. The wastegate is added, and the entire assembly is bolted to the rotary's engine block.

3. The turbo oil drain hose is connected from the bottom of the turbocharger to the previously installed fitting in the oil pan.

4. A banjo bolt in the engine block just above the oil pan is removed. Its head is

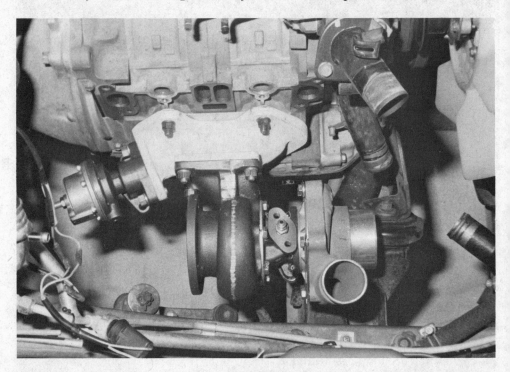

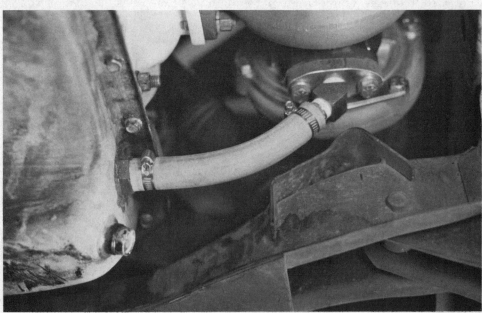

drilled and tapped, and it is reinstalled. The oil pressure hose's fitting is screwed into the banjo bolt's newly threaded hole, and the other end of the hose is connected to the oil fitting in the turbocharger.

5. The intake manifold is installed, and the throttle cable attached. On goes the brake boost bleed fitting,. oil metering valve cable, and the vacuum/boost bleed.

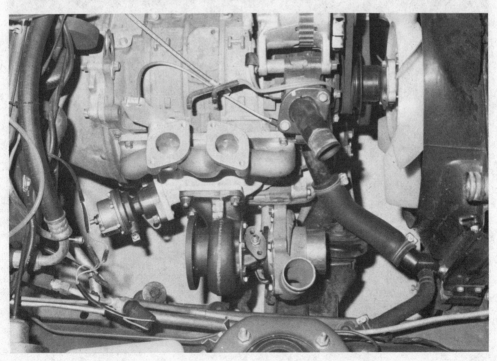

6. The carburetor is bolted onto the intake manifold, and the oil metering lines are attached to the carb. The oil metering valve cable assembly is connected to its linkage, and the cable is adjusted. The front engine lift hook is drilled, and the return spring end bolt is installed. The return spring is slipped into place.

The throttle cable is shortened slightly at the pedal end by cutting the pivot ball off and sliding on a new clamping cable end. The throttle cable is fine—adjusted with the adjusting nuts at the carburetor, and the throttle pedal stop (under the pedal) is adjusted to prevent straining the cable at full throttle. The cruise control cable is attached, and adjusted by removing the control unit's cover and shortening the cable inside. The choke cable is attached to the carburetor, and the brake booster vacuum bleed hose is connected to the fitting in the intake manifold.

7. The fuel pressure regulator is mounted to the inner fender, and rubber fuel line connects the metal fuel line to the regulator, and the regulator to the carb.

8. The new fuel pump is installed. The stock spring clamps used for fuel line connections are replaced with aircraft—quality hose clamps.

9. After a complete exhaust system is installed, it's time to add the boost gauge. The gauge's vacuum/pressure line is routed through a grommet in the firewall to a fitting in the intake manifold. The distributor vacuum line goes to the vacuum bleed on the carburetor. The crankcase vent, oil filler tube vent, and fuel tank vent are spliced together with "T"-fittings and routed to the vacuum canister.

10. The air–to–water intercooler's radiator and water pump are bolted onto their mounts, and this assembly is then hose clamped onto the cross tube in front of the engine coolant radiator.

11. The windshield washer bottle is mounted on the right inner fender behind the strut tower, and the charcoal canister moves to the old windshield washer bottle location. The radiator overflow bottle is mounted on the left strut tower.

12. The intercooler pump switch is connected to a manifold pressure hose and its electrical circuit is completed. When boost pressure is present, the switch provides the ground for the intercooler pump so it runs only when necessary.

13. The intercooler reservoir is mounted on the right strut tower, just above the level of the intercooler element located in the carb hood. The air filter assembly is installed.

14. The carb hood is installed, and the intercooler coolant hose system is completed. The intercooler system is filled with coolant, and its operation checked by connecting a jumper wire across the switch terminals. The anti–surge by-pass switch is installed in the vacuum bleed system. Engine oil and coolant are added.

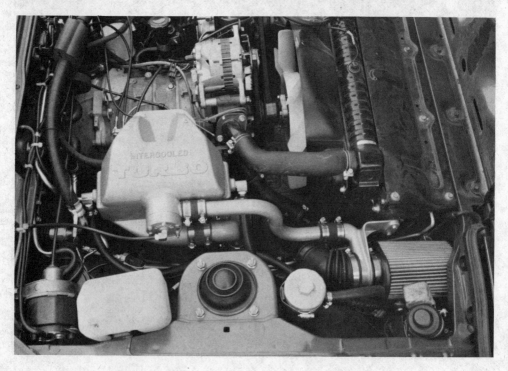

15. With the oil pressure line loose and ignition system disabled, the engine is cranked until oil flows out of the line. The pressure line's fitting is then tightened, and the engine is cranked for ten more seconds to assure thorough turbocharger oiling. Finally it's time for engine start-up, and the carburetor idle mixture, idle speed, and ignition timing are adjusted. During the test run, CarTech recommends reaching normal operating temperature without using any boost, and then checking all operating functions (oil pressure, etc.). Then boost pressure can be applied and increased in one psi increments, checking for detonation each time.

7

Two Wheeled Turbos

Motorcycles are the ultimate value in performance machinery. For $4000, one can buy a Japanese bike off the showroom floor which will accelerate from a standstill to 140 miles per hour in eighteen seconds: one half the time of the quickest production Porsche, Ferrari, or Corvette. These motorcycles will stop so short that care must be used to avoid being rear-ended by a car. Of course a bike's maneuverability is something else, and 40 miles per gallon (not EPA, but actual mileage) on regular gasoline puts the icing on the cake.

Yet the turbo, that ultimate performance elixir, has never had a firm hold on the motorcycle world. In 1975, American Turbo-Pak introduced a turbo kit for the Kawasaki Z-1, which made what was then the world's most powerful motorcycle 25% more powerful in street trim. Several turbo kit manufacturers have offered kits for various bikes—primarily big Japanese street bikes—since then, but none have caused more than a ripple of interest to the mass of performance-oriented motorcycle enthusiasts.

In 1982, Honda introduced the CX500 Turbo, and it was not well-received. It suffered from severe throttle lag and mediocre power when not under boost, but when the turbo finally kicked in, the CX500 Turbo would take off. This behavior is treacherous on a motorcycle. When swooping though a curve at speed, a rider must be able to apply power smoothly and at precisely the right moment (within a couple of hundredths of a second). Throttle lag and poor throttle response can cause a car to go into a slide near the limit of adhesion: on a bike, they cause certain disaster. The CX500 Turbo gave relatively poor 34 mpg gas mileage, cost much more, weighed more, and was slower than the 750cc normally-aspirated street bikes.

Ironically, the high-tech Honda Turbo is based on the only Honda motorcycle engine to use the antediluvian pushrod valvetrain system since the early '60s. However, Honda felt the CX500's watercooling was critical to turbo engine durability. Honda strengthened the V-twin CX's engine for turbo duty by upgrading its plain bearings and substituting forged 7.2:1 compression ratio pistons instead of the normally-aspirated version's 10:1 cast pistons. A strengthened clutch sends the power to higher ratio heat treated transmission gears. A larger finned sump keeps the oil cool.

An IHI turbocharger pumps the CX500 Turbo up to an incredible 17.4 pounds of boost. The turbocharger sits in front of the engine. A surge tank is located between the turbocharger and intake manifolds, damping the uneven firing V-twin's intake pulsations. In addition, a resonance chamber reduces intake resonances. Port fuel injection is used, necessitating a fuel pump and pressure regulator. The air filter is mounted behind the headlight and the fuel injection control unit is under the seat's tail section. A separate ignition control unit in the tail section adjusts the ignition timing on the basis of boost pressure and rpm.

Honda CX650 Turbo.

For 1983, Honda dropped the 500 and introduced the CX650 Turbo, which actually displaced 674ccs. The CX650 is a huge improvement over the 500, offering superb mid-range power, reduced throttle lag, and better fuel economy. The compression ratio was bumped up to 7.8:1, the intake valves are four millimeters larger. Intake valve lift is up .5 millimeters. The airbox and intake tract are enlarged, as is the compressor wheel. The ignition control unit is integrated into the fuel injection computer. An ig-

Honda uses an IHI turbocharger and port fuel injection on the V-twin.

nition sensor, an air pressure sensor, and the resonance chamber are eliminated. Maximum boost pressure is down one psi to 16.4 psi. The 500's fiberglass fairing is replaced by one of ABS plastic on the 650. A taller fifth gear improves fuel economy and ups the top speed by eleven miles per hour over the CX500 Turbo. And the CX650 Turbo weighs nine pounds less.

A few months after Honda introduced the CX500 Turbo, Yamaha gave us the XJ650LJ Seca Turbo. Based on their normally-aspirated air cooled four in-line DOHC two valve per cylinder 650 engine, the Yamaha turbo received 8.5:1 forged pistons and graphite-coated rings. The connecting rods have oil splash holes to cool the undersides of the pistons, and there is an oil cooler. A 750 Seca clutch is used. Mitsubishi's smallest turbocharger has a 39 millimeter diameter turbine, in comparison with the Honda's 48 millimeter wheel. Stainless steel-lined headers lead the exhaust gases to the turbocharger, located behind the engine, below the swing arm pivot. (The liners help retain exhaust gas energy for the turbo.) The left muffler handles normal exhaust flow while the right muffler only quiets the wastegate's exhaust flow.

The Yamaha's Mitsubishi turbocharger blows through four Mikuni CV carburetors. A camshaft-driven fuel pump keeps the carburetor bowls full. Maximum boost is 8 psi. A reed valve-equipped compressor bypass passage improves engine breathing when not under boost. When the turbo's compressor is turning slowly enough, the compressor and housing actually serve as an intake restriction. Under these conditions, the reed valve in the bypass passage opens and allows air to enter under minimal restriction. As boost pressure takes over, the reed valve closes again so the turbo will have maximum effect. An ignition computer sets the ignition timing on the basis of vacuum/boost and rpm, and retards the timing step by step when the detonation sensor hears the sound of detonation.

The Yamaha is the slowest of the factory turbobikes. Its throttle response and fuel economy are fairly good. Yet a number of normally aspirated 650s are quicker, get better fuel economy, offer perfect throttle response, are easier to maintain, and cost a couple of thousand dollars less.

The Yamaha 650 Seca Turbo's Mitsubishi turbocharger blows through four carburetors.

Suzuki XN85 Turbo.

For 1983, Suzuki introduced their air-cooled XN85 Turbo. Displacing 674 ccs, this engine is based on Suzuki's normally aspirated, air cooled DOHC two valve per cylinder 650 in-line four. The connecting rods, crankpins, cylinder liners, cylinder studs and wrist pins are beefed up for turbo duty. Metal head and base gaskets replace the fiber gaskets used on the normally aspirated engine. 7.4:1 compression ratio pistons are used (instead of the standard 9.5:1 version), and both the top and second rings are chromed, as opposed to just the top ring on the normally-aspirated version.

Four oil jets squirt oil at the undersides of the pistons at high rpm—a check valve cuts off flow at low rpm so as not to overcool. A new cylinder head casting flows oil around the exhaust valve seats for cooling, and grooves in the exhaust tappet cavities improve oil flow. The plain bearings of Suzuki's normally-aspirated shaft-drive model are used instead of the roller bearings of the chain drive version, for better reliability

The Suzuki has an IHI turbocharger and port fuel injection.

at lower oil pressures. An oil cooler, larger oil pump, and 500cc more oil capacity complete the oil system.

The IHI turbocharger sits behind the cylinders pumping a maximum of 9.8 pounds of boost through the intake manifold. The ported fuel injection system is made by Nippondenso, and the fuel pump and air flow meter are the same as those used by Toyota. Information concerning air flow, engine speed, engine and air temperature, barometric pressure, throttle position, and starter motor engagement (for starting enrichment) is sent to the seat tail-mounted fuel injection computer. An ignition computer resides beside the fuel injection computer, and this ignition computer incorporates boost-activated ignition retard. Fuel injection and ignition are momentarily cut off if the engine exceeds 10,000 rpm or is overboosted due to wastegate failure. The camshafts and valves are the same as on the normally-aspirated version.

The Suzuki XN85's engine performance puts it slightly behind the Honda but well ahead of the Yamaha Turbo. The Suzuki's off-boost to on-boost transitions are reasonably smooth. Its riding position is racer-like—less than comfortable for extended trips. While its handling is dead stable at high speeds, bumpy low speed curves upset the Suzuki.

At the end of 1983, Kawasaki released the fastest production turbobike ever, the GPZ 750 Turbo. Despite 30% less engine displacement, the GPZ 750 Turbo is as quick as an 1100 superbike. Yet the GPZ 750 Turbo's engine changes and turbo systems are relatively simple. Like the Suzuki and Yamaha Turbos, the Kawasaki is a DOHC two valve per cylinder four in-line: 7.8:1 compression ratio pistons replace the normally-aspirated version's 9.5:1 pistons. The combustion chambers have no squish bands. Camshaft lift is reduced by one millimeter, intake valve duration is 32 degrees shorter, and exhaust valve duration is 26 degrees shorter. This flattens the torque curve. The clutch has one more plate and a reinforced hub. The turbo has a larger diameter transmission mainshaft, larger transmission bearings, a sharper undercut on the gear dogs (for more secure engagement), a stronger primary drive chain, and different transmission ratios. A spring-loaded pre-transmission jackshaft damper replaces the normally-aspirated GPZ's rubber damper.

The Hitachi HT-10B turbocharger is mounted in front of the engine. The air filter sits beside the transmission. The turbocharger draws the air from the air cleaner, com-

The Kawasaki GPZ 750 Turbo is the fastest production turbo bike.

presses it to a maximum of 10.8 pounds of boost, and forces it through a steel crossover tube to a plenum chamber behind the engine.

The ported fuel injection system meters fuel on the basis of information concerning engine speed, throttle position, intake air temperature, engine temperature, and boost pressure. When the throttle is yanked open, an extra dose of fuel is squirted into the ports, the injection system acting as an accelerator pump to eliminate bog. The duration (and hence the amount) of this single pulse varies with the rate of throttle opening and the initial throttle position. The fuel injectors shut off to protect the engine when rpm exceeds 11,500 or boost exceeds 12.6 psi.

To avoid detonation, Kawasaki simply and crudely retarded maximum ignition advance by 14 degrees. The addition of a boost-only activated ignition retard system, such as the Arkay ADC, in conjunction with advancing the base ignition timing setting, would improve off-boost power, throttle response, and fuel economy. Ultimate turbo fans could also integrate an air-to-water intercooler into the GPZ 750 Turbo's plenum chamber. The intercooler's pump could be triggered by the ADC's water injection output wire. Such an installation might make an interesting project.

Talk about exciting projects, there's nothing that will get the blood of a performance enthusiast pumping faster than a ride on a turbo-kitted superbike. Mr. Turbo makes turbo kits for the Honda CB750/900/1100F, CBX, and Gold Wing, Yamaha XS11/Maxim, Suzuki 750/1000/1100, and Kawasaki 550/750/900/1000 and 1300. Their typical in-line four cylinder kit features chromed four-into-one tubular headers leading to a chromed Rajay turbocharger. An adjustable wastegate mounts to the exhaust header. A high exhaust pipe is standard, with a low pipe and muffler baffle optional. A round chrome air filter mounts to the Keihin butterfly carburetor. The cast aluminum plenum chamber routes the compressed fuel/air mixture to the intake ports. A boost gauge, electric fuel pump, oil lines, and hardware are all included.

While these kits can be bolted onto a stock engine with low boost, low compression pistons are highly recommended, and Mr. Turbo carries the superb MTC piston line.

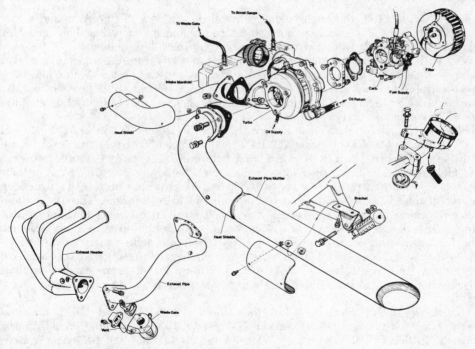

Layout of a typical Mr. Turbo kit on an in-line four cylinder.

Mr. Turbo-kitted Kawasaki KZ1000.

A street Suzuki 1100, thoroughly equipped with heavy duty engine parts and a Mr. Turbo kit, put out 180 rearwheel horsepower on Cycle magazine's dynomometer. This bike was subsequently ridden coast to coast with no reliability problems.

Mr. Turbo also offers racing versions of their turbo kits featuring a higher flow fuel pump, 30 psi boost gauge, and modified wastegate and carburetor. Water injection, boost-activated ignition retard systems, and deep sump oil pans are also available. Mr. Turbo's Gold Wing kit has a water heated intake manifold for better cold driveability as well as water injection.

Blake Enterprises makes turbo kits for the Honda CB750/900 and Gold Wing, and CBX, Kawaskai 900/1000, Suzuki 750/1000/1100, Yamaha 750 triple and XS11. Some of Blake's kit's strong points include minimal throttle lag, excellent throttle response and fuel economy, making their kits excellent for all-around street use. Most of their kits use chromed short tube or log-type headers which retain heat better than long individual tube headers, at the expense of optimum high rpm flow. The die cast aluminum plenum chambers are carefully designed for even cylinder-to-cylinder distribution. The slide-type Mikuni carburetor requires no fuel pump in most applications. The Rajay turbocharger is controlled by Blake's own adjustable wastegate, the BPM, which uses a stainless steel Manley valve. An optional compressor bypass is available for minimal throttle lag. Low exhaust pipes and chrome air filters are standard on most kits. Racing options include water injection, larger Mikunis and a special turbine housing.

RB Racing offers turbo kits for: the Honda CB750/900 and CBX; Kawasaki 550/650/750/900/1000/and 1100; Suzuki 550/650/750/1000/ and 1100; and Yamaha 550/650/750 Secas, 750/850 triples, 750/920 Virago, and 750/1100. RB Racing's basic street kits use IHI turbochargers with integral wastegates and a chromed low exhaust

A Mr. Turbo-kitted Honda Gold Wing makes the ultimate tourer.

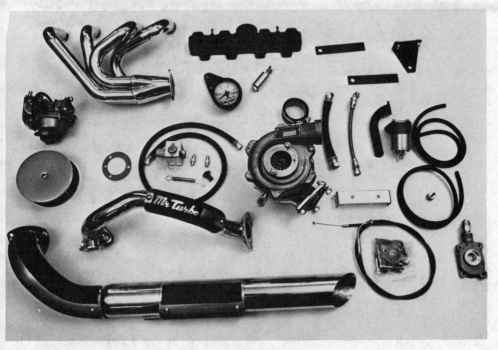

Mr. Turbo kits are very complete.

Blake's kit for the Kawasaki KZ1000.

Suzuki GS1100 fitted with a Blake turbo kit.

system which loops to the front of the bike before making its way to the rear (which includes a removable baffle). The modified constant velocity SU carburetor features an accelerator pump. Chromed four-into-one headers lead to the turbocharger. A fuel pump with built in adjustable pressure regulator, boost gauge, and braided oil lines are all included. RB Racing's stage II systems use a conventional turbo-style exhaust system (with removable baffle) and substitute a header-mounted Boost Guard wastegate for the IHI's integral unit. The Boost Guard can be used with an optional Dial-A-Boost maximum boost pressure adjuster. Twin turbo kits are available for the Honda

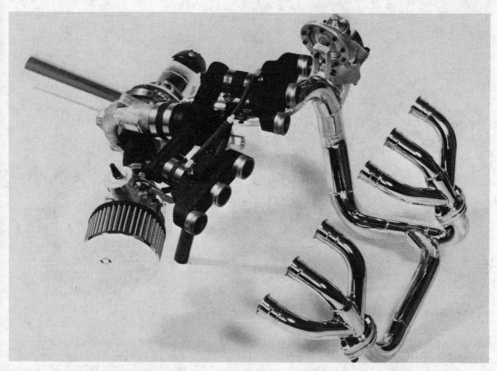

Blake's Honda CBX six cylinder kit.

RB Racing sells this Honda CBX turbo kit—twin turbo kits are also available.

A Luftmeister turbo-powered BMW.

CBX, Suzuki 1100, and Kawasaki 1000/1100. RB Racing also has water injection kits, a tell-tale boost gauge (retains the maximum boost figure until a release button is pressed), electric turbo bearing oil pump/dry sump systems for the Suzuki 750/1000 and Kawasaki 900/1000/1100, and a deep sump/heavy duty oil pump system for the big Kawasakis. Luftmeister has a series of turbo kits for the BMW boxers. An IHI RH05 turbocharger is used with an integral wastegate. The Keihin carburetor features an accelerator pump. The air cleaner and exhaust system are black chromed. A Lockhart

Luftmeister's kit even includes an oil cooler and three-pawl clutch.

oil cooler with braided lines, boost gauge, battery heat shield, hardened cam follower buttons, heavy duty valve springs, and a three-pawl clutch are all included.

While turbo kits are available for many bikes, the motorcycle turbo aftermarket has not been able to attract mass interest, and sighting a turbo-kitted bike on the street is a rare occurrence. Similarly, sales of factory turbo bikes have been so disappointing that Honda, Suzuki, and Yamaha have discontinued their turbo bikes for '84, and are trying to sell out their leftover stock. Only the Kawasaki GPZ 750 turbo, the only factory turbo bike which can compete performance-wise with the 1100 superbikes, is available as an '84. This is ironic, since the Suzuki, Yamaha, and Honda could have all slipped into the U.S. without paying the ITC (International Trade Commission) over-700cc tariff—a penalty the Kawasaki turbo buyer must pay. But while the present turbo-bike scene lacks the excitement of the automotive turbo industry, who knows what the future will bring . . . maybe factory intercooled, electronically-wastegated turbo 1100s!

Aftermarket Turbo Equipment Manufacturers

Ak Miller
9236 Bermudez St.
Pico Rivera, CA 90660
(213) 949–2548

Arkay
14009 S. Crenshaw Blvd.
Hawthorne, CA 90250
(213) 675–9161

Gale Banks Engineering
546 Duggan Ave.
Azusa, CA 91702
(213) 969–9605

BAE
3032 Kashiwa St.
Torrance, CA 90505
(213) 530–4743

Blake Enterprises
Rt. 1 Box 403
Muskogee, OK 74401
(918) 683–2967

Bob Sharp Racing
21 South Street
Danbury, CT 06810
(203) 743–4487

Callaway Turbosystems
3 High Street
Old Lyme, CT 06371
(203) 434–9002

Car Tech
11144 Ables Lane
Dallas, TX 75229
(214) 620–0389

Corvette America
Box 427, Route 322
Boalsburg, PA 16827
(814) 364–1435

DM Engineering
612 Federal Road
Brookfield, CT 06804
(203) 775–1853

IPD
2762 N.E. Broadway
Portland, OR 97232
(503) 287–1179

Jafco
3016 6-B Halladay Street
Santa Ana, CA 92705
(714) 641–5881

Larson Engineering
26121 Van Born Road
Taylor, MI 48180
(313) 292–6643

Luftmeister
135 Stanley St.
Compton, CA 90220
(213) 638–8508

Martin Turbo
1310 Johnson Drive
City of Industry, CA 91745
(213) 965–0781

Mr. Turbo
8002 South Madison St.
Burr Ridge, IL 60521
(312) 986–5669

Pfaff Turbo
2152 O'Toole Ave.
San Jose, CA 95131
(408) 946–9751

RB Racing
9063 W. Washington Blvd.
Culver City, CA 90230
(213) 838–6464

Martin Schneider Designed Systems
9063 W. Washington Blvd.
Culver City, CA 90230
(213) 559–0020

Shankle Automotive Engineering
9135-F Alabama Ave.
Chatsworth, CA 91311
(213) 709–6155

Spearco Performance Products
7541 Woodman Place
Van Nuys, CA 91405
(213) 901–7851

Turbocharger
12215 S. Woodruff, Box 687
Downey, CA 90241
(213) 773–1880

Turbo International
7091-A Belgrave
Garden Grove, CA 92641
(714) 891–5704

Turbo Toms
4090 Peachtree Rd. NE
Atlanta, GA 30319
(404) 458–5055

Turbo Terminology

A/R ratio—the ratio between the turbine nozzle area and the distance from the center of the turbine wheel to the centroid of the nozzle area. This ratio affects the turbine speed for a given exhaust flow, a larger ratio causing the turbine to spin slower.

blow through—a turbocharger system in which the turbocharger blows air through the carburetor(s) or fuel injector(s), i.e., the fuel and air mixing occurs downstream from the turbocharger.

boost-activated ignition retard—a system which retards the ignition timing when the intake manifold is under pressure, in order to reduce the chance of detonation.

boost gauge—a meter that measures boost pressure

boost pressure—the pressure of compressed air or fuel/air mixture in the intake manifold.

combustion chamber—the cavity in the cylinder head (or cylinder head and piston) into which the fuel/air mixture is compressed by the piston when the piston is at the top of its compression stroke.

compression ratio—the ratio between the volume in the cylinder when the piston is at the bottom of its stroke and the volume in the cylinder when the piston is at the top of its stroke.

compressor—the section of a turbocharger that compresses the intake air or fuel/air mixture.

compressor pressure ratio—the ratio between the absolute pressure at the compressor outlet and the absolute pressure at the compressor inlet.

density—the weight of air or fuel/air mixture per unit volume.

detonation—the uncontrolled spontaneous explosion of fuel/air mixture in the combustion chamber, causing a loss of power and possible engine damage.

detonation-activated ignition retard—a system which retards the ignition timing when the detonation sensor picks up vibration at frequencies typical of detonation.

detonation sensor—a device, usually piezoelectric, which senses frequencies typical of detonation and converts this information into an electric current.

displacement—the total volume of an engine's cylinder(s), usually measured in cubic inches, cubic centimeters, or liters. One liter equals 1000 cubic centimeters, which equals 61 cubic inches.

draw through—a turbocharger system in which the turbocharger sucks the fuel/air mixture through the carburetor or fuel injector, i.e., the fuel and air mixing occurs upstream from the turbocharger.

electronically-controlled wastegate—a wastegate that is activated by an electric signal from a computer.

enrichment system—an auxillary fuel injection system designed to add additional fuel to the intake mixture only under boost conditions.

exducer—the outermost section of a turbine wheel, used to purge the turbine of exhaust gases.

horsepower—one horsepower is the amount of power required to lift 550 pounds one foot per second.

ignition timing—the moment at which the sparkplug fires, usually expressed in the number of crankshaft degrees before the piston reaches the top of its stroke.

inducer—the section of the compressor wheel that draws the air or fuel/air mixture into the compressor.

intercooler—a radiator used to reduce the temperature of the compressed air or fuel/air mixture before it enters the combustion chamber(s).

lean fuel/air mixture—a greater proportion of air in the fuel/air mixture than the stoichiometric mixture.

mean effective pressure—the average pressure in the cylinder during the power stroke minus work lost in compressing the air or fuel/air mixture.

normally aspirated—an engine which draws its fuel/air mixture into its cylinder(s) solely by piston-created vacuum, i.e., not turbocharged or supercharged.

oil cooler—a radiator which cools oil, generally engine oil, but sometimes transmission or differential oil.

oxygen sensor—a device fitted to an engine's exhaust manifold which senses the oxygen content in the exhaust and converts this information into an electric current.

preignition—the premature burning of fuel/air mixture in the combustion chamber caused by combustion chamber heat and/or fuel instability. Preignition begins before the spark plug fires.

pyrometer—an exhaust gas temperature indicator.

rich fuel/air mixture—a greater proportion of fuel in the fuel/air mixture than the stoichiometric mixture.

rotor—the rotating assembly of a turbocharger, including the compressor wheel, shaft, and turbine wheel.

rpm—engine speed, measured in crankshaft revolutions per minute.

stoichiometric mixture—the proper ratio of air to fuel for complete and efficient combustion, around 15:1 by weight for gasoline engines.

supercharger—a device which forces intake air or fuel/air mixture into an engine's cylinder(s) under pressure and which is driven mechanically, usually via a belt and pulley from the engine's crankshaft.

surge line—the line on a compressor map which repesents the minimum stable flow at each pressure ratio, below which compressor output is unsteady.

turbine—the section of a turbocharger that converts exhaust gas energy into rotary motion.

turbocharger—a device which forces intake air or fuel/air mixture into an engine's cylinder(s) under pressure and which is driven by the engine's exhaust gases.

volumetric efficiency—the ratio between the weight of fuel/air mixture actually drawn into a cylinder under operation and the weight of fuel/air mixture which woould fill that cylinder at atmospheric pressure and temperature.

wastegate—a device which bleeds off exhaust gases before they reach the turbocharger when boost pressure reaches a set limit.

water injection—a system which injects water (and/or alcohol) into an engine's intake air to reduce detonation.